Surveys and Tutorials in the Applied Mathematical Sciences

Volume 19

Series Editors

Anthony Bloch, Department of Mathematics, University of Michigan, Ann Arbor, MI, USA

Charles L. Epstein, Department of Mathematics, University of Pennsylvania, Philadelphia, PA, USA

Alain Goriely, Mathematics, University of Oxford, Oxford, UK

L. Greengard, Department of Mathematics, Courant Institute of Mathematical Sciences, New York University, New York, USA

Advisory Editors

Michael Brenner, School of Engineering, Harvard University, Cambridge, USA

Gábor Csányi, Engineering Laboratory, University of Cambridge, Cambridge, UK

Lakshminarayanan Mahadevan, School of Engineering, Harvard University, Cambridge, USA

Clarence Rowley, MAE Department, Princeton University, Princeton, USA

Amit Singer, Fine Hall, Princeton University, Princeton, USA

Jonathan D. Victor, Weill Cornell Medical College, New York, NY, USA

Rachel Ward, Department of Mathematics, University of Texas at Austin, Austin, TX, USA

AF670334

Featuring short books of approximately 80-200pp, Surveys and Tutorials in the Applied Mathematical Sciences (STAMS) focuses on emerging topics, with an emphasis on emerging mathematical and computational techniques that are proving relevant in the physical, biological sciences and social sciences. STAMS also includes expository texts describing innovative applications or recent developments in more classical mathematical and computational methods.

This series is aimed at graduate students and researchers across the mathematical sciences. Contributions are intended to be accessible to a broad audience, featuring clear exposition, a lively tutorial style, and pointers to the literature for further study. In some cases a volume can serve as a preliminary version of a fuller and more comprehensive book.

Marcelo Epstein

An Elementary Introduction to n-fold Groupoids

Contextualized by Elastic Composites and Metamaterials

Springer

Marcelo Epstein
Department of Mechanical and
Manufacturing Engineering
SSE University of Calgary
Calgary, AB, Canada

ISSN 2199-4765 ISSN 2199-4773 (electronic)
Surveys and Tutorials in the Applied Mathematical Sciences
ISBN 978-981-95-1868-5 ISBN 978-981-95-1869-2 (eBook)
https://doi.org/10.1007/978-981-95-1869-2

Mathematics Subject Classification: 18B40, 18D05, 74E30, 74A40, 20N02, 74B20

© The Editor(s) (if applicable) and The Author(s), under exclusive license to Springer Nature Singapore Pte Ltd. 2026

This work is subject to copyright. All rights are solely and exclusively licensed by the Publisher, whether the whole or part of the material is concerned, specifically the rights of translation, reprinting, reuse of illustrations, recitation, broadcasting, reproduction on microfilms or in any other physical way, and transmission or information storage and retrieval, electronic adaptation, computer software, or by similar or dissimilar methodology now known or hereafter developed.
The use of general descriptive names, registered names, trademarks, service marks, etc. in this publication does not imply, even in the absence of a specific statement, that such names are exempt from the relevant protective laws and regulations and therefore free for general use.
The publisher, the authors and the editors are safe to assume that the advice and information in this book are believed to be true and accurate at the date of publication. Neither the publisher nor the authors or the editors give a warranty, expressed or implied, with respect to the material contained herein or for any errors or omissions that may have been made. The publisher remains neutral with regard to jurisdictional claims in published maps and institutional affiliations.

This Springer imprint is published by the registered company Springer Nature Singapore Pte Ltd.
The registered company address is: 152 Beach Road, #21-01/04 Gateway East, Singapore 189721, Singapore

If disposing of this product, please recycle the paper.

A la querida memoria de
esa superpareja única,
Noé y Catalina, mis padres

Preface

In his celebrated 1964 ‘lost lecture’ Richard Feynman said:

> I am going to give what I will call an elementary demonstration. But elementary does not mean easy to understand. Elementary means that very little is required to know ahead of time in order to understand it, except to have an infinite amount of intelligence.

The last eight words are, of course, typical of Feynman’s naughty sense of humour. But the spirit of his premise is that any person would be able to understand his lecture (on planetary motion) if only he or she were driven by curiosity and interest in sufficient amount to overcome the sometimes arduous task of following the details of a logical argument. He went on to promise that not even an acquaintance with calculus would be required to understand how he derives the exact shape of the trajectory of a planet about the sun in the absence of any other mass. Newton himself did not obtain his solution by integrating a system of differential equations, as done today in textbooks. Instead he used a Greek-style geometric setting, not completely unlike Feynman’s manner of presentation.

The notions of groupoid and n-fold groupoid are non-trivial extensions of the mathematical concept of group. Groups appear in physics as manifestations of the general idea of symmetry. Groupoids can be regarded as a generalization that, in the words of the mathematician Alan Weinstein [34], provide a unifying framework to consolidate the internal and external symmetries, or local and distant symmetries, of a system. Further, n-fold groupoids permit us to conceive symmetries in the space of symmetries, some sort of symmetries between symmetries. Just as in the case of Feynman’s lost lecture, the gist of these notions can be explained in terms that, at least in principle, do not require any previous mathematical background.

The mechanics of deformable solids, with its sensorial concreteness, provides a fertile ground for the application of these ideas. Indeed, a point in a solid body may be made of a material endowed with symmetries. For example, untreated steel in a natural state has the same properties in all directions. We can regard this property as a local rotational symmetry. But when comparing two different points of the body we may ask the question: are these points made of the same material? If one is able to imagine (or actually perform) a transplant of a small neighbourhood of the first point into a neighbourhood of the second in such a way that the graft cannot be distinguished from the original material thereat by any experiment, we may say that we have discovered a distant symmetry between these points, and the concept of groupoid arises in a natural way.

The topic of local and distant symmetries is admittedly not as exciting as Feynman's in his lost lecture. For who is not interested in the millennia-old topic of planetary motion? On the other hand, if you are interested in acquiring a working familiarity with groupoids and n-fold groupoids, and if you search the literature, you may find that what Feynman described as 'what is required to know ahead of time in order to understand' will be dauntingly insufficient to delve into the existing treatises. As far as ordinary groupoids are concerned, corresponding to the case $n = 1$ in the book title, there are many good sources available. The classical treatises of Kyril Mackenzie [22, 24] emphasize the differential-geometric aspects of the theory, while Ronald Brown's [5] avenue is mainly topological. A gentler treatment can be found in the book by Ibort and Rodríguez [20], although their treatment is quite technical and does not discuss the application to solid mechanics. A book that is devoted to this application, by de León et al. [8], is far too advanced for a first encounter with the topic. The present book can be considered an elementary introduction to groupoids in general and to this application in particular. Another peculiarity of our presentation is that it includes the much less explored and more difficult conceptual framework of n-fold groupoids and their application to solid mixtures. Even for these more sophisticated structures, the present treatment does not assume any prerequisites, not even calculus of several variables. Some sections have been indicated with a star to warn the reader about the appearance of slightly more advanced material.

In essence, this book is about symmetries. A symmetry is an operation that leaves a physical or mathematical entity unchanged. If we rotate a pentagon by 72 degrees about its centre, we recover an exact copy of the original figure. This is the kind of symmetry embodied in the mathematical concept of *group*. But we may also consider symmetries of exchange between parts of a system. For example, if two identical tiles are removed from their placements in the kitchen floor and transplanted so that they replace one another, we recover an exact copy of the original floor. This is a kind of *distant symmetry*, not encompassed under the concept of group. The notion of *groupoid*, which made its first appearance in mathematics in 1927 [4], generalizes the notion of group in a way that allows for the incorporation of these distant symmetries. It was later found out that groupoids are particular cases of *categories*. Among several further generalizations in the domain of categories we find n-fold (or multiple) categories [9, 16], a particular case of which are n-fold groupoids.

It is a remarkable feature of continuum mechanics, that is, the mechanics of deformable media, that all these abstract notions turn out to have a concrete counterpart within the mechanical context. If symmetries are interpreted as operations that leave some feature of the material response unchanged, groups, groupoids, and n-fold groupoids find a comfortable home in contexts such as continuous distributions of dislocations, composites, mixtures, and metamaterials. These contextualizations are confined mainly to Chaps. 3 and 5, and a small portion of Chap. 7. They should not impair the independent reading of the remaining chapters that provide an elementary introduction to the conceptual structure and the terminology of the theory of n-fold groupoids.

In writing this short book it has not been my intention, even if it had been within the compass of my abilities, to make original contributions to groupoid theory or to 3D printing, additive manufacturing or metamaterials. My aim has been to provide a fresh and light-hearted approach to the somewhat unexpected connection between an area as abstract as category theory and another apparently very distant area of important technological applications. My hope is that it may inspire others to pursue a better blend of theory and practice. Twenty centuries ago, speaking about the ideal architect, Vitruvius already knew that

> Those who without sufficient education turn their hand to practice are not able to succeed in attaining influence for their works, while those who rely exclusively on theory and education seem to be pursuing not a thing but a mere shadow of reality.

Calgary, AB, Canada
2025

Marcelo Epstein

Acknowledgment I wish to express my thanks to Professor Samer Adeeb, at the University of Alberta, for his critical reading of an early version of the manuscript.
M. E.

Competing Interests The author has no competing interests to declare that are relevant to the content of this manuscript.

Contents

Chapter 1
What Is a Groupoid?

Abstract The collection of all symmetries of rotation and reflection of a plane figure is used as an example to motivate a review of the concept of group, from its abstract definition to the consideration of other notions such as subgroups and conjugacy. A system consisting of several plane figures is then exploited to demonstrate that in addition to the individual symmetries of each of the figures (which may be called local symmetries) it may be worth considering also distant symmetries of congruence between two or more of these figures. The incorporation of both local and distant symmetries under a single mathematical umbrella leads to the notion of groupoid, whose precise definition in a general setting that transcends these examples is provided in detail and illustrated with other, not necessarily geometric, examples. A groupoid can be visualized as a set of arrows hovering over a set of objects, called the base of the groupoid.

1.1 Local Symmetries and Groups

The five-pointed star represented in Fig. 1.1 has rotational symmetries. What do we mean by that exactly? If we rotate the star by 72° counter-clockwise about its centre, we recover the same figure. Since this is the case, we conclude that a further rotation of 72° will also leave the star in congruence with its initial configuration, except that each ray will now appear in a new position (for example, a_0 has now moved to the position previously occupied by a_2).

This intuitive feeling for the notion of symmetry of a figure can be given a slightly more formal character. If we denote by r_1 the counter-clockwise 72-degree rotation, we can refer to the successive application of two such rotations as a *composition*, which we will denote by $r_1\,r_1$. As indicated before, this composition results in a counter-clockwise rotation of 144°, which we may call r_2, and it turns out that r_2 is also a symmetry (namely, an operation that exactly reproduces the initial star). Consider now the successive application of r_1 followed by r_2. We will notate this composition by $r_2\,r_1$. By convention, we will agree that the rotation that was applied

© The Author(s), under exclusive license to Springer Nature Singapore Pte Ltd. 2026
M. Epstein, *An Elementary Introduction to n-fold Groupoids*,
Surveys and Tutorials in the Applied Mathematical Sciences 19,
https://doi.org/10.1007/978-981-95-1869-2_1

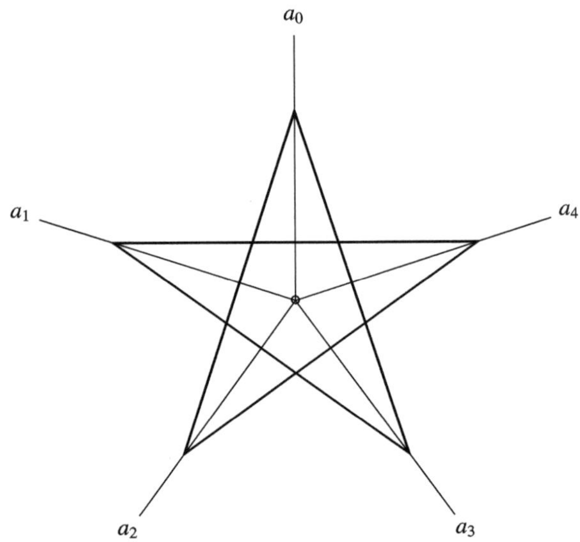

Fig. 1.1 A five-pointed star

first appears as the second factor of the composition. Clearly, this composition too is a rotational symmetry, say r_3.

How many such distinct rotational symmetries can we obtain in this way? If we agree to identify a rotation of 360° with the trivial zero rotation r_0, we obtain a total of five distinct rotations, namely, r_0, r_1, r_2, r_3, r_4, where the subscript indicates the number of times the 72-degree rotation has been applied. We notice also that each member r_k (with $k = 0, \ldots, 4$) of this collection has an *inverse* element, which we denote by r_k^{-1}. The inverse element reverses the effect of r_k, in the sense that $r_k r_k^{-1} = r_k^{-1} r_k = r_0$. The inverse of each element is a member of the collection. Indeed, we can easily verify that

$$r_0^{-1} = r_0, \quad r_1^{-1} = r_4, \quad r_2^{-1} = r_3, \quad r_3^{-1} = r_2, \quad r_4^{-1} = r_1. \tag{1.1}$$

We will call the collection $\{r_0, r_1, r_2, r_3, r_4\}$ a *group* with *unit* (or *identity*) r_0. In our example, we say that this group *acts* on the figure producing *transformations* of the figure into itself. The identity element of the group is also called its *neutral element* since it leaves everything unaltered.

A peculiarity of this example is that the composition operation is indifferent to the order of the factors, namely, for any indices k and h we obtain $r_k r_h = r_h r_k$. We refer to this property by saying that the group is *commutative* or *Abelian*. It is not difficult to produce an example of a *non-commutative* group. We note that the star of Fig. 1.1 has also symmetries of reflection. Indeed, if you imagine a two-sided mirror aligned with, say, the ray a_0, the rays a_1 and a_4 will exchange positions, and so will the rays a_2 and a_3, while the ray a_0 will remain unchanged. If we denote by f_k the reflection about a_k, the collection $\{r_0, r_1, r_2, r_3, r_4, f_0, f_1, f_2, f_3, f_4\}$ also forms a group with identity element r_0. The inverse of f_k is f_k itself (for each

$k = 0, 1, \ldots 4$). This group, however, is non-commutative. For example, we can easily verify that

$$f_0 r_1 \neq r_1 f_0. \tag{1.2}$$

Exercise 1.1 Verify Eq. (1.1).

Exercise 1.2 Starting from Fig. 1.1 show that $f_0 r_1 \neq r_1 f_0$ by comparing the final positions of the rays $a_0, \ldots, a_4$.

Before introducing the formal definition of a group (which is essentially a generalization of the examples discussed above), let us review some mathematical notations which come handy when reading more technical treatments. Without questioning the logical foundations of set theory, we will agree to consider a *set* B as an arbitrary collection of objects, called the *elements* of B. If an object b is an element of B, we say that *b belongs to B*, and we indicate this as $b \in B$. If the object b is not an element of B we write $b \notin B$. The symbol $\emptyset$ stands for the *empty set*, the set with no elements. A set with just one element is called a *singleton*, written as $\{b\}$. We remark that the element b is not the same as the set $\{b\}$. The symbol $\forall\, b$ stands for the expression 'for all b', and the combination of symbols $\exists\, b\ :$ stands for 'there is (or exists) some b such that'. The arrow $\Rightarrow$ stands for 'implies that'. A set A is said to be a *subset* of another set B if $b \in A \Rightarrow b \in B$. We write this as $A \subset B$.

Given two sets, B and C, the *Cartesian product* $B \times C$ is the set of all *ordered pairs* (p, q), where $p \in B$ and $q \in C$. Clearly, unless B and C coincide, we have

$$B \times C \neq C \times B. \tag{1.3}$$

The real line $\mathbb{R}$ is the set of all real numbers. One way to visualize this set geometrically is by identifying it with an oriented straight line. Choosing a unit of measurement, and one point of the line as *origin*, the correspondence between real numbers (in, say, a decimal representation) and points is established. Given two real numbers $a < b$ the segment $[a, b] \subset \mathbb{R}$ consists of all points $x \in \mathbb{R}$ such that $a \leq x \leq b$. It is also called a *closed interval* of the real line. The product $\mathbb{R}^2 = \mathbb{R} \times \mathbb{R}$ consists of all the ordered pairs of real numbers. It can be identified with the Cartesian plane with coordinates x, y.

Exercise 1.3 Show or explain the geometrical meaning of each of the following sets: $\{a\} \times \{b\}$, $\{a\} \times [b, c]$, $[1, 2] \times [1, 3]$, $[1, 3] \times [1, 2]$, $[0, 1]^3$, where a, b, c are real numbers with $b < c$, and where $[0, 1]^3 = [0, 1] \times [0, 1] \times [0, 1]$.

A *function* $f : B \to C$ from a set B to a set C assigns to each element b of B an element $f(b)$ of C. Functions are also called *maps*, and we say that the function f maps the elements of B into elements of C. The element $f(b) \in C$ is also called the *image* of b. A function is said to be *injective* or *one-to-one* if no two elements of B are mapped into the same element of C. A function is said to be *surjective* or *onto* if every element of C is the image of at least one element of B. A function which is both injective and surjective (that is, one-to-one and onto) is called a *bijection* of A

with B. Given a bijection $f : A \to B$ we can uniquely define the *inverse function* $f^{-1} : B \to A$, which is also a bijection of B with A. The inverse function f^{-1} assigns to each element $b \in B$ the (unique) element $a \in A$ such that $f(a) = b$. The *identity map* of a set A is the trivial bijection $\mathrm{id}_A : A \to A$ such that $\mathrm{id}_A(a) = a$ for all $a \in A$.

If two functions $f : A \to B$ and $g : B \to C$ are given, their *composition* is the function $g \circ f : A \to C$ that assigns to each element $a \in A$ the element $g(f(a)) \in C$. Note, incidentally, that the function f is applied first to deliver $f(a) \in B$, and only then g is applied to this result. This explains the somewhat counter-intuitive notation, whereby the second operator appears to the left of the first.

Definition 1.1 A *group* consists of a set $\mathcal{G}$ and a function

$$f : \mathcal{G} \times \mathcal{G} \to \mathcal{G} \tag{1.4}$$

called the group *product* or *multiplication* or *composition* which is usually denoted by simple apposition, that is,

$$g_1\, g_2 = f(g_1, g_2), \qquad \forall g_1, g_2 \in \mathcal{G}. \tag{1.5}$$

The group product satisfies the following properties:

1. *Associativity*

$$(g_1\, g_2)\, g_3 = g_1\, (g_2\, g_3), \qquad \forall g_1,\ g_2,\ g_3 \in \mathcal{G}. \tag{1.6}$$

2. *Identity element*

$$\exists\, e \in \mathcal{G} : e\, g = g\, e = g, \qquad \forall g \in \mathcal{G}. \tag{1.7}$$

3. *Inverse elements*

$$\forall g \in \mathcal{G}\ \exists\, g^{-1} \in \mathcal{G} : g\, g^{-1} = g^{-1}\, g = e. \tag{1.8}$$

Exercise 1.4 Verify that the statements (1.6), (1.7) and (1.8) can be worded (somewhat awkwardly) as follows:

1. Given the product $g_1\, g_2\, g_3$ of three elements g_1, g_2 and g_3, as long as we do not alter their order, we may multiply the first two elements and then multiply the result by the third, and obtain the same result as if we had multiplied the first element by the product of the last two elements.
2. There exists a neutral element (or identity) e such that the result of multiplying it on the left or on the right by any element g of the group is g itself.
3. For every group element g there is a corresponding inverse element g^{-1} such that the result of multiplying it on the right or on the left by g is the identity element.

Exercise 1.5 Prove that the identity and the inverses are unique. [Hint: assume there are two different identity elements and calculate their product.]

Exercise 1.6 The symmetry group of our five-pointed star consists of a finite number of elements. Show that: (a) the set $\mathbb{Z}$ of all integer numbers constitutes an infinite commutative group under the operation of addition; (b) the set $\mathbb{Q}+$ of positive rational numbers under multiplication is a commutative group; (c) the set $\mathbb{Q}-$ of negative rational numbers under multiplication is not a group. These sets are countably infinite,[1] but there are also examples of *continuous groups*, which are very important in physical applications. A simple example is the collection of all the rotational symmetries of a circle. Another example is the collection $\mathbb{R} - \{0\}$ of all non-zero real numbers with the operation of multiplication.

Definition 1.2 A *subgroup* of a group $\mathcal{G}$ is a subset $\mathcal{H} \subset \mathcal{G}$ which is a group under the same group operation of $\mathcal{G}$.

We may further clarify this definition by saying that the subset $\mathcal{H}$ is closed under the operations of composition and inversion. The smallest possible subgroup of a group $\mathcal{G}$ is the *trivial subgroup* consisting of the singleton $\{e\}$.

Example 1.1 The collection of all bijections $f : A \to A$ of a set A onto itself constitutes a group whose neutral element is the identity map and whose operation is the composition of functions. When A is a set with a finite number n of elements, this group is known as the *symmetric group of degree* n, denoted by S_n. In this case, it is convenient to identify the set A with the ordered n-tuple

$$A = \{1, 2, \ldots, n\}, \tag{1.9}$$

and to regard each bijection as a *permutation* $\sigma : A \to A$, that is, a re-ordering of the entries of the n-tuple. For example, for $n = 6$ a permutation σ may be defined as

$$\sigma(\{1, 2, 3, 4, 5, 6\}) = \{3, 2, 6, 5, 1, 4\}. \tag{1.10}$$

The repeated application of this σ would produce

$$\sigma(\sigma(\{1, 2, 3, 4, 5, 6\})) = \sigma(\{3, 2, 6, 5, 1, 4\}) = \{6, 2, 4, 1, 3, 5\}. \tag{1.11}$$

The inverse permutation σ^{-1} in our example is

$$\sigma^{-1}(\{1, 2, 3, 4, 5, 6\}) = \{5, 2, 1, 6, 4, 3\}. \tag{1.12}$$

[1] A set is countably infinite if it can be put in a one-to-one correspondence with the natural numbers. The set $\mathbb{Q}$ of rational numbers is countable, while the set $\mathbb{R}$ of all real numbers is not. Intuitively, this means that 'there are more reals than fractions'.

A *transposition* is a permutation that leaves all elements unaltered except two elements that exchange their values. A common notation for the transposition of the entries with values i and j is (i, j). It can be shown that every permutation can be obtained (non-uniquely) as a successive composition of a number of transpositions. The *parity* (even or odd) of the number of transpositions is independent of the succession chosen. Accordingly, the *sign* or *signature* of a permutation is defined as $+1$ for an even permutation and -1 for an odd permutation.

Exercise 1.7 Verify Eqs. (1.11) and (1.12).

Exercise 1.8 Verify that the permutation σ in Eq. (1.10) can be obtained as

$$\sigma = (1, 3) \circ (3, 5) \circ (3, 6) \circ (4, 6), \tag{1.13}$$

and also as

$$\sigma = (4, 5) \circ (5, 6) \circ (1, 2) \circ (2, 3) \circ (3, 4) \circ (4, 5) \circ (3, 4) \circ (1, 2). \tag{1.14}$$

Recall that we are always adhering to the reverse order of compositions, whereby the rightmost function (in this case a transposition) is applied first.

An important concept in group theory is that of *conjugacy*. Two elements, g_1 and g_2 of a group $\mathcal{G}$ are said to be *conjugate* if there is an element $h \in \mathcal{G}$ such that

$$g_2 = h\, g_1\, h^{-1}. \tag{1.15}$$

Exercise 1.9 Show that conjugacy is an *equivalence relation*, namely: (i) every group element is conjugate to itself; (ii) if g_2 is conjugate to g_1, then g_1 is conjugate to g_2; (iii) if g_1 is conjugate to g_2, and g_2 is conjugate to g_3, then g_1 is conjugate to g_3. These properties characterizing an equivalence relation are known, respectively, as *reflexivity*, *symmetry*, and *transitivity*.

As an equivalence relation, conjugacy partitions the group $\mathcal{G}$ into disjoint (i.e., non-intersecting) *equivalence classes* known as *conjugacy classes*. The (unique) conjugacy class to which an element $a \in \mathcal{G}$ belongs is the set G_a defined as

$$G_a = \{g \in \mathcal{G} \,:\, \exists h \in \mathcal{G} \,:\, g = h\, a\, h^{-1}\}. \tag{1.16}$$

In simpler words, the conjugacy class G_a is the subset of $\mathcal{G}$ comprising all the elements conjugate to a.

Exercise 1.10 Find all the conjugacy classes of: (a) the group of rotations of the five-pointed star; (b) the group of rotations and reflections of the five-pointed star.

Exercise 1.11 Show that the conjugacy class of the identity e of a group is the singleton $G_e = \{e\}$. Prove that in a commutative group all conjugacy classes are singletons, and vice versa.

If $\mathcal{H} \subset \mathcal{G}$ is a subgroup of $\mathcal{G}$, and if $a \in \mathcal{G}$, the set

$$\mathcal{H}_a = a\mathcal{H}a^{-1} = \{g \in \mathcal{G} \ : \ g = aha^{-1} \text{ for some } h \in \mathcal{H}\} \tag{1.17}$$

is also a subgroup of $\mathcal{G}$. It is called the *conjugate subgroup of* $\mathcal{H}$ *by* a.

Exercise 1.12 Prove that a conjugate of a subgroup as defined above is indeed a subgroup.

Exercise 1.13 Prove that the map $\mathcal{H} \to \mathcal{H}_a$ given by Eq. (1.17) is a bijection. It may be said, therefore, that conjugate subgroups represent the same concept as seen from two different points of view, or two different frames of reference.

1.2 Distant Symmetries and Groupoids

Looking now at an example on a grander scale, like a beach or a meadow $\mathcal{B}$, we perceive several planar objects of various shapes, as shown in Fig. 1.2. Each of these objects, considered on its own, has a (*local*) group of symmetries (rotational, say) which can be as small as the trivial group (consisting of the neutral element alone) or as large as the continuous group of rotational symmetries of the circle. The total collection of objects, however, does not appear to have any non-trivial rotational symmetry. Intuitively, on the other hand, we realize that the three stars in Fig. 1.2 are mutually related by what we may call *distant symmetries*. Similarly, the two circles and two of the triangles are related by distant symmetries. These symmetries consist of individual translations and rotations that would bring two distant objects into coincidence. The notion of *groupoid* is a generalization of the notion of group that allows for the consolidation of both kinds of symmetry under a single umbrella.

Given an ordered pair (X, Y) of objects, X and Y, of the set $\mathcal{B}$ we can imagine an *arrow* u with *source* at X and *target* at Y for every distant symmetry (if any) that would map X into Y. We will denote the source of an arrow u by $\alpha(u)$ and its target by $\beta(u)$. If Y coincides with X we obtain a loop-shaped arrow. Obviously, these loop-shaped arrows represent the local symmetries! At each object X there

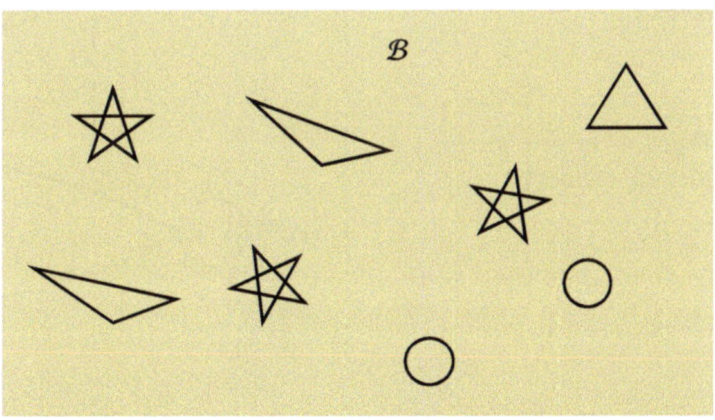

Fig. 1.2 Objects on a meadow

is at least one such arrow, $\epsilon(X)$, representing the always available trivial symmetry of the object, namely, the identity of the local symmetry group. The collection of loop-shaped arrows at X is a graphical representation of the symmetry group of the object X.

In this way, for any given set of objects $\mathcal{B}$ (in which a concept of symmetry between objects has been given some physical or geometrical meaning) we obtain a corresponding set of arrows $\mathcal{Z}$. For any non-empty set of objects $\mathcal{B}$, the set $\mathcal{Z}$ is never empty, since it contains at least the loop-shaped arrows representing the identity elements of the local symmetry groups of each object. Given an arrow $u \in \mathcal{Z}$, as already indicated, it has a source (or tail) $\alpha(u) \in \mathcal{B}$ and a target (or tip) $\beta(u) \in \mathcal{B}$. In other words, we have two maps

$$\alpha : \mathcal{Z} \to \mathcal{B}, \qquad \beta : \mathcal{Z} \to \mathcal{B}, \tag{1.18}$$

which are called, respectively, the *source projection map* and the *target projection map*.

Exercise 1.14 Show that the projection maps are surjective. [Hint: use an argument based on the loop-shaped arrows.]

The close connection between the set of objects and the set of arrows will be indicated by the suggestive notation $\mathcal{Z} \rightrightarrows \mathcal{B}$, the two arrows hinting at the two projection maps. But there is more. We can define an operation of *composition* between arrows. This operation is based on the intuitive observation that if an object X has a distant symmetry u with an object Y, and if Y has a distant symmetry v with W, then X surely must have a distant symmetry with W. More graphically, two arrows can be composed if, and only if, the tip of u coincides with the tail of v. We denote by $v\,u$ the resulting arrow with tail at X and tip at W, as shown in Fig. 1.3.

For each arrow u from X to Y there is a corresponding *inverse arrow* u^{-1} from Y to X, as shown in Fig. 1.4. These two arrows are related by

$$u^{-1}\,u = \epsilon(X) \tag{1.19}$$

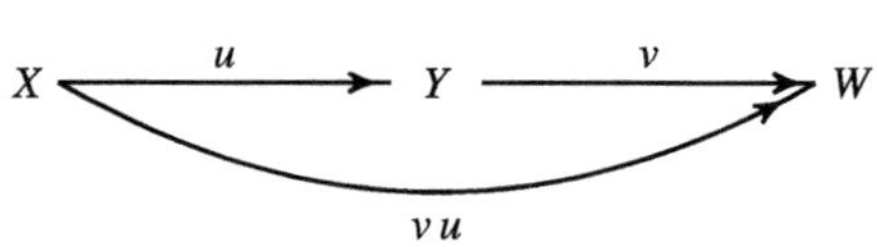

Fig. 1.3 Composition of arrows

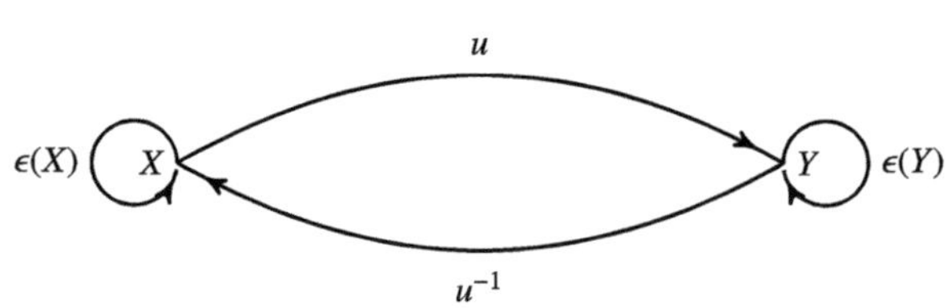

Fig. 1.4 Inverse arrows and identity elements

and

$$u\,u^{-1} = \epsilon(Y). \tag{1.20}$$

With these rules of composition, $\mathcal{Z} \rightrightarrows \mathcal{B}$ becomes a *groupoid* with *base* $\mathcal{B}$ and source and target projections α and β. The term *groupoid* alludes to the fact that, just as in a group, we have a set $\mathcal{Z}$ and a binary operation (composition) defined among its elements (the arrows). This operation, however, is not defined for every pair of elements but only for those pairs that satisfy an extra condition (namely, the tip-to-tail connection). The operation is associative and inverses exist and are unique. Instead of a single identity, we have now an identity arrow attached to each element of an underlying set $\mathcal{B}$ of objects onto which the set $\mathcal{Z}$ has two projections. A group is a particular case of a groupoid that corresponds to the set of objects $\mathcal{B}$ being a singleton (that is, a set with just one element).

Exercise 1.15 Consider two stars, X and Y say, in Fig. 1.2. Each of these stars has exactly five rotational symmetries. Show that the number of distant rotational symmetries (arrows) from X to Y is also necessarily five. [Hint: consider two such arrows, u and v, and form the composition $v^{-1}\,u$.]

Definition 1.3 A *groupoid* $\mathcal{Z} \rightrightarrows \mathcal{B}$ consists of a set $\mathcal{Z}$ (the set of arrows) and a set $\mathcal{B}$ called the *base* of the groupoid (the set of objects). These sets are related by means of two surjective *projection maps* $\alpha : \mathcal{Z} \to \mathcal{B}$ and $\beta : \mathcal{Z} \to \mathcal{B}$ called, respectively, the *source map* and the *target map*. A binary internal operation (*product* or *composition*) between two arrows $u, v \in \mathcal{Z}$ is defined whenever the 'tip-to-tail' condition

$$\beta(u) = \alpha(v) \tag{1.21}$$

is satisfied. The result, denoted by vu, is an arrow such that

$$\alpha(vu) = \alpha(u), \qquad \beta(vu) = \beta(v). \tag{1.22}$$

This operation, whenever possible, is associative. For each object $X \in \mathcal{B}$ a (unique) *identity arrow* $\epsilon(X) \in \mathcal{Z}$ exists such that

$$u\epsilon(X) = u, \qquad \forall u \in \mathcal{Z} : \alpha(u) = X \tag{1.23}$$

and

$$\epsilon(X)u = u, \qquad \forall u \in \mathcal{Z} : \beta(u) = X. \tag{1.24}$$

Finally, for each $u \in \mathcal{Z}$ there is an *inverse arrow* $u^{-1} \in \mathcal{Z}$ such that

$$u^{-1}u = \epsilon(\alpha(u)), \qquad u\,u^{-1} = \epsilon(\beta(u)). \tag{1.25}$$

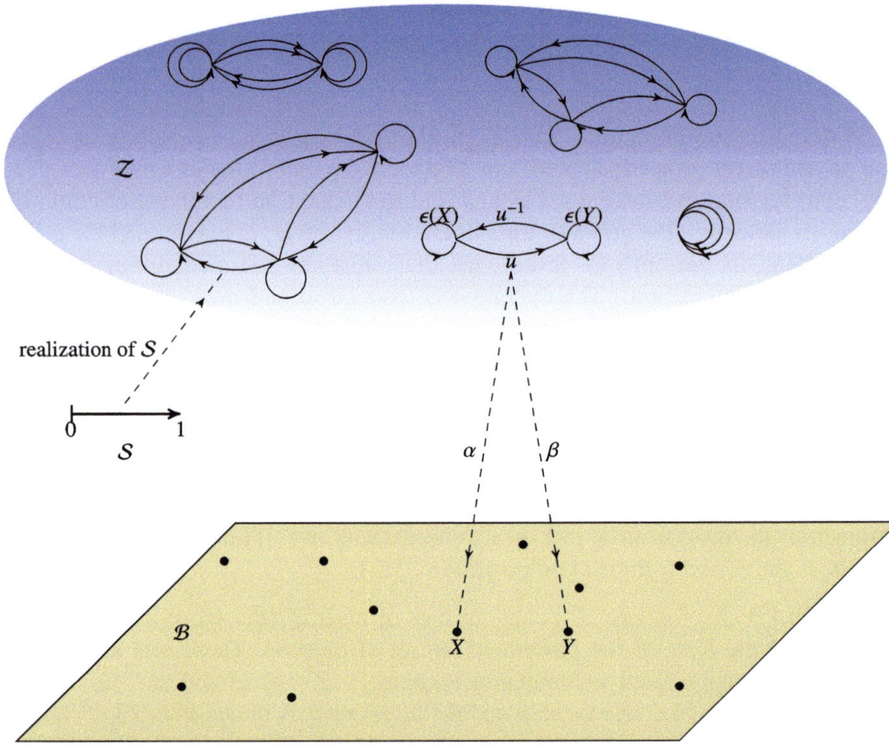

Fig. 1.5 A groupoid $\mathcal{Z}$ as a cloud of arrows hovering over a meadow $\mathcal{B}$ with objects

Remark 1.1 An artistic rendering of a groupoid is shown in Fig. 1.5. It is important to bear in mind that, in principle, neither the objects nor the arrows of a groupoid are made of material points. In fact, the 'arrow' u only symbolizes the fact that between the objects $\alpha(u)$ and $\beta(u)$ there is a *morphism*, that is, a map whose physical or mathematical meaning has been well-defined in some context of interest. A point of view that may be useful later is to consider an abstract ideal arrow such as the unit oriented segment $\mathcal{S} = [0, 1] \in \mathbb{R}$. Then, given a groupoid $\mathcal{Z} \rightrightarrows \mathcal{B}$, every arrow in $\mathcal{Z}$ is a *realization* of this ideal arrow $\mathcal{S}$. The origin of $\mathcal{S}$ corresponds to the tail of the groupoid arrow. Remember that these pictorial representations have only a motivational value and are not to be considered substitutes for precise definitions.

Exercise 1.16 Show that the (never empty) collection of all arrows u such that $\alpha(u) = \beta(u) = X$ constitutes a group $\mathcal{G}_X$ (known as the *vertex group* or the *symmetry group* at $X \in \mathcal{B}$).

Exercise 1.17 Show that if u is an arrow with $\alpha(u) = X$ and $\beta(u) = Y$, the collection $\mathcal{P}_{XY}$ of all arrows with source X and target Y is exactly given by

$$\mathcal{P}_{XY} = u\,\mathcal{G}_X = \mathcal{G}_Y\,u. \tag{1.26}$$

In words, the vertex groups at an object are in a one-to-one correspondence with the arrows (if any) issuing from it to another object, or arriving at it from another object.

Exercise 1.18 Show that if u is an arrow with $\alpha(u) = X$ and $\beta(u) = Y$, the vertex groups at X and Y are *conjugate* in the sense that

$$\mathcal{G}_Y = u\,\mathcal{G}_X\,u^{-1}. \tag{1.27}$$

Definition 1.4 A groupoid $\mathcal{Z} \rightrightarrows \mathcal{B}$ is said to be *transitive* if there is at least one arrow between every pair of objects.

Clearly, according to Exercise 1.18, in a transitive groupoid all the vertex groups are mutually conjugate. The converse is not true. The vertex groups of a groupoid may be conjugate without the groupoid being transitive.

Exercise 1.19 Let $\mathcal{Z} \rightrightarrows \mathcal{B}$ be a groupoid. The map $[\alpha, \beta] : \mathcal{Z} \to \mathcal{B} \times \mathcal{B}$ given by $[\alpha, \beta](u) = \{\alpha(u), \beta(u)\}$ is known as the *anchor map* of the groupoid. It assigns to each arrow u the ordered pair of points of $\mathcal{B}$ consisting of the tail and the tip of u. Show that a groupoid is transitive if, and only if, the anchor map is surjective.

Example 1.2 Let the object set $\mathcal{B}$ consist of three rectangles, shown in Fig. 1.6a. Considering only rotational congruences, we obtain a transitive groupoid whose arrows are pictured in Fig. 1.6b.

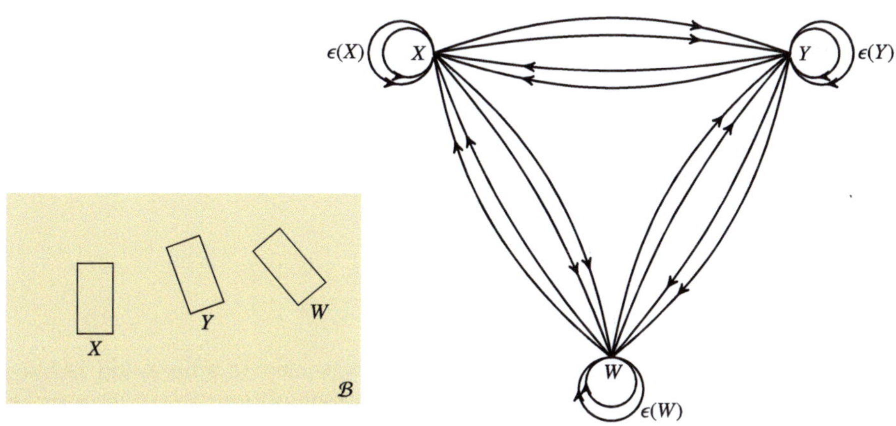

(*a*) (*b*)

Fig. 1.6 A transitive groupoid

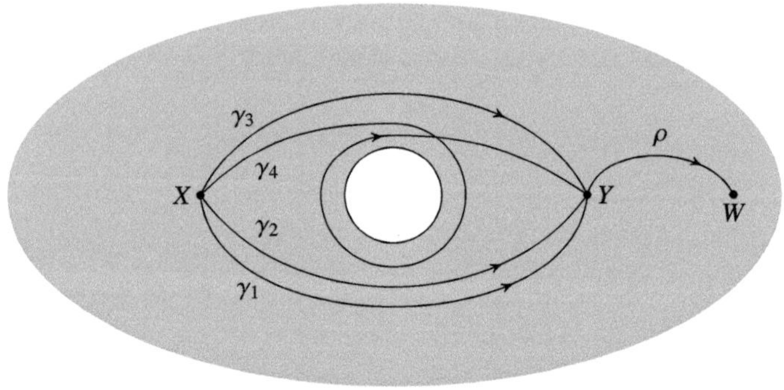

Fig. 1.7 A punctured plane

Example 1.3 (The Pair Groupoid) A simple example of a groupoid is the so-called *pair groupoid* of a set $\mathcal{B}$. The arrows of this groupoid are the elements (b, c) of $\mathcal{B} \times \mathcal{B}$. The projection maps are defined as

$$\alpha((b, c)) = b, \qquad \beta((b, c)) = c. \tag{1.28}$$

The composition (or product) and inverse are given by

$$(b, c)\,(a, b) = (a, c), \qquad (b, c)^{-1} = (c, b). \tag{1.29}$$

We have adhered to the convention that the second factor in a product is applied first. The identity arrow at $b \in \mathcal{B}$ is $\epsilon(b) = (b, b)$. We emphasize that the meaning of the groupoid arrows in each particular example depends entirely on the context.

Example 1.4 (*The Fundamental Topological Groupoid) Consider as a set of objects $\mathcal{B}$ a plane with a hole, or *punctured plane*, as shown in Fig. 1.7. Given two points $X, Y \in \mathcal{B}$, a *path* from X (the source) to Y (the target) is defined as a continuous curve joining these two points, that is, a continuous map $\gamma : [0, 1] \to \mathcal{B}$ such that $\gamma(0) = X$ and $\gamma(1) = Y$. If $X = Y$ the path can be called a *loop*. Given a path γ from X to Y and a path ρ from Y to W, we can define the *concatenation* $\gamma * \rho$ as a path from X to W defined as

$$\gamma * \rho(s) = \begin{cases} \gamma(2s) & \text{if } 0 \le s \le 1/2 \\ \rho(1 - 2s) & \text{if } 1/2 < s \le 1 \end{cases}. \tag{1.30}$$

Intuitively, $\gamma * \rho$ is just the combination of the two paths in a tip-to-tail fashion. Notice that the concatenation is not exactly the composition $\rho \circ \gamma$. Two paths, γ_1, γ_2, between X and Y are said to be *homotopic* if they can be continuously deformed into each other while keeping the end points fixed. Homotopy is, clearly, an equivalence relation. Accordingly, the collection of all curves between X and Y can be partitioned into disjoint equivalence classes called *homotopy classes*. These classes can be concatenated in an obvious manner by choosing a representative path from each class. Therefore, we can construct a groupoid $\mathcal{Z} \rightrightarrows \mathcal{B}$ (known as the *fundamental groupoid of* $\mathcal{B}$), where the elements of $\mathcal{Z}$ are the homotopy classes. The identity at X is the homotopy class of the constant curve, $\epsilon(s) = X$. In the context of the punctured plane, even without the paraphernalia of the formulas above, we can intuitively see that the curves γ_1 and γ_2 in Fig. 1.7 are homotopic, but γ_3 is not homotopic to either of them. Moreover, γ_4 is not homotopic to γ_3, because it involves an extra (clockwise) revolution about the hole. As a result of these considerations (without providing a rigorous proof) we can conclude that the homotopy classes (and so the elements of $\mathcal{Z}$) are as many as the integers $\mathbb{Z}$.

Exercise 1.20 (* Path-connectedness and transitivity) Suggest an intuitive argument to justify the following proposition: The fundamental groupoid of $\mathcal{B}$ is transitive if, and only if, $\mathcal{B}$ is path-connected. By path-connectedness we mean that for any two points $X, Y \in \mathcal{B}$ there is a path with source X and target Y.

Definition 1.5 A groupoid $\mathcal{Z} \rightrightarrows \mathcal{B}$ is *totally intransitive* if $\mathcal{Z}$ is the union of all the vertex groups. In other words, all the arrows of the groupoid are loop-shaped.

Exercise 1.21 Show that a groupoid is totally intransitive if, and only if, the image of the anchor map is the set of the identities of the pair groupoid of the base.

1.2.1 Subgroupoids

Informally speaking, a *subgroupoid* $\mathcal{U} \rightrightarrows \mathcal{A}$ of a groupoid $\mathcal{Z} \rightrightarrows \mathcal{B}$ consists of subsets $\mathcal{U} \subset \mathcal{Z}$ and $\mathcal{A} \subset \mathcal{B}$ such that $\mathcal{U} \rightrightarrows \mathcal{A}$ is closed under the restriction of the operations of composition and inversion, and under the source and target projections. There are two extreme cases that deserve special attention. A subgroupoid is said to be *full* if $\mathcal{A}$ is a proper subset of $\mathcal{B}$ while all the arrows between the remaining elements are preserved. In other words, the set of objects is shrunk, but nothing else changes as far as the distant symmetries are concerned.

Example 1.5 A simple illustration of a full subgroupoid can be obtained if in our Example 1.2 we were to disregard the rectangle W. We would still obtain a transitive groupoid whose graphical representation is depicted in Fig. 1.8.

At the other extreme, a subgroupoid is said to be *wide* if $\mathcal{A} = \mathcal{B}$. In other words, the set of objects remains unchanged, but there is a possible reduction in the set of arrows. This is a case of significant meaning in applications. In the case of a transitive groupoid, as we already know, all the vertex groups are mutually conjugate, and we may speak of a typical vertex group $\mathcal{G}$. A wide subgroupoid may consist in a reduction of the typical group to a subgroup $\mathcal{H} \subset \mathcal{G}$. To illustrate this situation in a very simple example, let us assume that in Example 1.2 we had included not only the rotational but also the reflection symmetries of the rectangles.

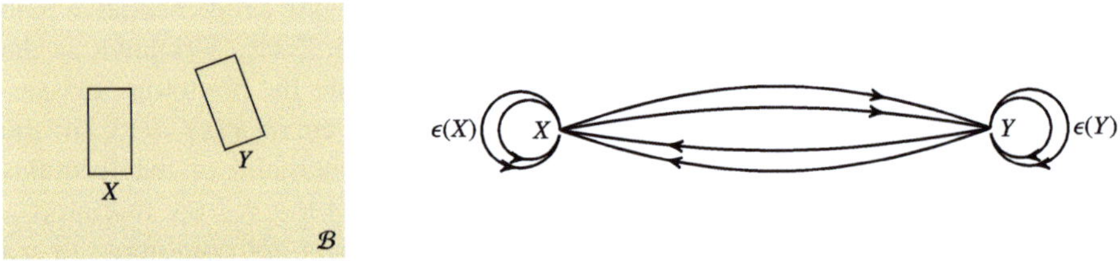

Fig. 1.8 A full subgroupoid of the groupoid of Fig. 1.6

Then, the groupoid of the example (in which reflection symmetries are ignored) would be a wide subgroupoid of the former.

Example 1.6 (The Base Subgroupoid) The *base subgroupoid* $\Delta_{\mathcal{Z}}$ of a groupoid $\mathcal{Z} \rightrightarrows \mathcal{B}$ is the wide subgroupoid obtained by retaining only the identity arrow at each point of the base $\mathcal{B}$. Any set $\mathcal{B}$ can be considered as a groupoid with this structure. The base subgroupoid is, by definition, totally intransitive.

In general, a subgroupoid need not be either wide or full. Consider the pair groupoid $\mathbb{R} \times \mathbb{R} \rightrightarrows \mathbb{R}$. Its arrows are ordered pairs of real numbers. The subgroupoid $\mathbb{Z} \times \mathbb{Z} \rightrightarrows \mathbb{Z}$, in which the arrows are pairs of integers, is a full subgroupoid of $\mathbb{R} \times \mathbb{R} \rightrightarrows \mathbb{R}$. The base subgroupoid $\Delta_{\mathbb{Z} \times \mathbb{Z}}$ is a wide subgroupoid of $\mathbb{Z} \times \mathbb{Z} \rightrightarrows \mathbb{Z}$, but it is neither full nor wide as a subgroupoid of $\mathbb{R} \times \mathbb{R} \rightrightarrows \mathbb{R}$.

Chapter 2
A Surgical-Mechanical Imagery

Abstract In the spirit of bringing into play the concrete contextualization alluded to in the book subtitle, the familiar medical procedure of skin grafting is invoked to suggest an instance in which the local and distant symmetries can be regarded as linear transformations. Thus, for example, a rectangular piece of skin harvested from the source tissue may have to be stretched and deformed into a parallelogram before grafting it at the target location. A review of elementary concepts in matrix algebra in two and three dimensions is undertaken. The general linear group and some of its important subgroups are defined for later use.

2.1 Skin Grafts

When skin is damaged due to a burn, an infection, or other causes, a common remediation policy is a surgical procedure known as a *skin graft* or *skin transplantation*. Healthy skin is harvested from an unaffected area of the patient's body and transplanted onto the damaged region. If the area to be covered is relatively large, many smaller pieces can be used as patches to obtain a larger assembly, or the technique of *skin graft meshing* [19] may be adopted. Figure 2.1 is a simplified illustration of this concept. Although we are not attempting a medically accurate description of the procedure, let us accept this figure as the basis for a suggestive imagery.

A detail that does not go unnoticed, even in this frankly amateurish picture, is that skin (as many other soft tissues in humans and animals) is usually in a state of stretch with respect to its natural configuration. This situation (also referred to as the presence of *residual stresses*) is manifest most dramatically in arteries. When a ring segment of arterial wall is cut, it opens up visibly [15], as shown in Fig. 2.2, as internal stresses are relaxed. A similar effect is observable in skin tissue [3, 37]. If a small circular patch of epidermis is cut from the body, its shape becomes that of an ellipse. In more general terms, a body may be made of the same material at all points, just like one would obtain by joining together a large number of small skin implants. But each implant may have undergone some kind of stretch before the

© The Author(s), under exclusive license to Springer Nature Singapore Pte Ltd. 2026

M. Epstein, *An Elementary Introduction to* n*-fold Groupoids*,
Surveys and Tutorials in the Applied Mathematical Sciences 19,
https://doi.org/10.1007/978-981-95-1869-2_2

Fig. 2.1 Schematic of skin graft meshing

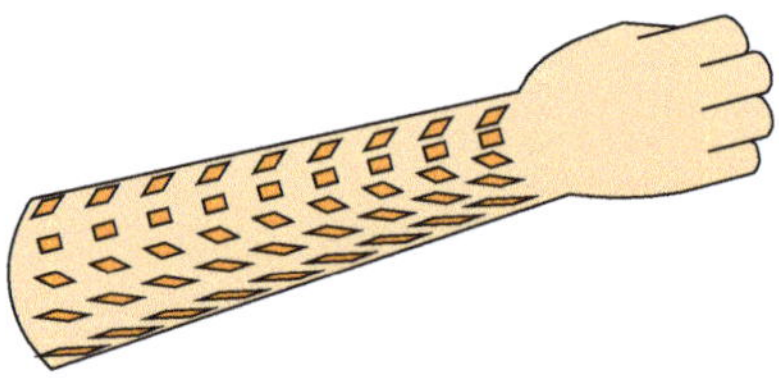

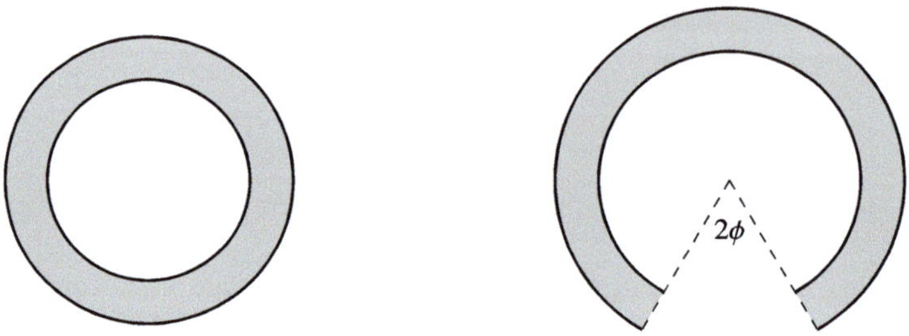

Fig. 2.2 Arterial ring before (left) and after (right) a radial cut

pieces were joined together when put in place. Accordingly, two distant pieces of the implanted skin can be made to come into exact correspondence with each other by a transplant operation that may involve some stretching. If this is the case, we will draw an arrow from the first to the second point to indicate that they are made of the same material. If the pieces are made of different materials, no such transplant operation will do the job of rendering them indistinguishable from each other, and we will draw no arrow connecting them.

2.2 The General Linear Group

In some of the examples presented in the previous chapter, the transplant operations boiled down to rigid rotations. But now we have considered a more general case, whereby the transplant includes also a stretch. To describe this operation in more precise mathematical terms, we will first exemplify it in a two-dimensional context.

A point P in the standard plane is located by means of its coordinates x_P and y_P. Let a, b, c, and d be some constants, and let the point P be moved to a new position P' with coordinates

$$x'_P = a\,x_P + b\,y_P, \qquad y'_P = c\,x_P + d\,y_P. \tag{2.1}$$

If we use the same formula (with the same values for the constants a, b, c, d) for the corners of a unit square $OPQR$ we obtain a parallelogram. Figure 2.3 displays the result for the arbitrarily chosen values $a = 0.8, b = 0.4, c = 0.2, d = 0.7$.

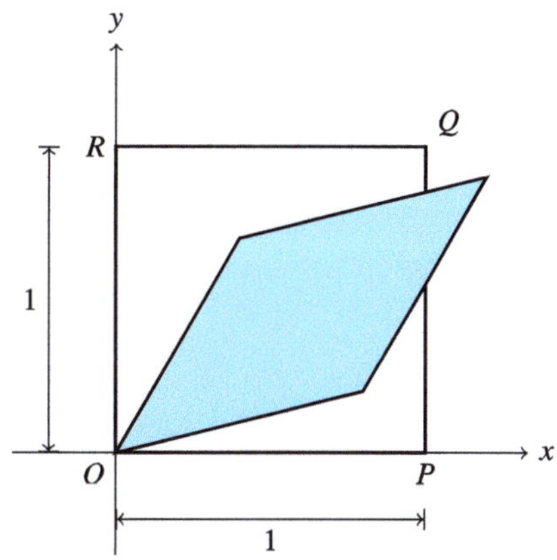

Fig. 2.3 Linear transformation of a square in its plane

This operation is called a *linear transformation*. We can represent it by arranging the constants in a square matrix A as

$$[A] = \begin{bmatrix} a & b \\ c & d \end{bmatrix} \tag{2.2}$$

and arrange the coordinates into two column vectors as

$$\{X\} = \begin{Bmatrix} x \\ y \end{Bmatrix}, \qquad \{X'\} = \begin{Bmatrix} x' \\ y' \end{Bmatrix}. \tag{2.3}$$

Equation (2.1) can be rewritten more compactly as

$$\{X'\} = [A]\{X\}, \tag{2.4}$$

where the operation of *matrix multiplication* has been used (according to which the elements of the m-th row of the first matrix are successively multiplied by the elements of the n-th column of the second and the results added together to produce the element in the m-th row and the n-th column of the product matrix, as shown in Example 2.1 below). If $[A]$ happens to be the *unit matrix* $[I]$, namely,

$$[I] = \begin{bmatrix} 1 & 0 \\ 0 & 1 \end{bmatrix}, \tag{2.5}$$

the points $\{X\}$ stay put in their original positions, and we say that the matrix $[I]$ represents the *identity transformation*. In this event, the square of Fig. 2.3 does not undergo any deformation.

Example 2.1 (Matrix Multiplication)

$$[A]\,[B] = \begin{bmatrix} a & b \\ c & d \end{bmatrix} \begin{bmatrix} e & f \\ g & h \end{bmatrix} = \begin{bmatrix} ae + bg & af + bh \\ ce + dg & cf + dh \end{bmatrix}. \tag{2.6}$$

The *determinant* of the square matrix $[A]$ is the number

$$\det(A) = a\,d - b\,c. \tag{2.7}$$

This number vanishes if, and only if, the entries in the second row are proportional to the corresponding elements in the first row, namely, if there exists a real number α such that $c = \alpha\,a$ and $d = \alpha\,b$. When this happens, the matrix is said to be *singular* and the parallelogram of Fig. 2.3 collapses into a line, a situation with a drastic physical meaning. This *dimensional collapse* is nicely described in [26] as the effect of stepping on an egg crate or a cardboard moving-box, thus losing one of the three dimensions. If the matrix has a negative determinant, the dimensionality is preserved but the orientation is reversed, just as when seeing our right hand reflected in a mirror.

Exercise 2.1 Choose any numerical value for the constant α and set $c = \alpha a, d = \alpha b$. Check that in Fig. 2.3 the parallelogram would collapse into a line.

When $\det(A) \neq 0$ the linear transformation can be *inverted*, by which we mean that an *inverse matrix* $[A]^{-1}$ can be obtained such that the parallelogram is transformed back into the square. The inverse matrix (in the two-dimensional case we are considering) can be easily obtained as

$$[A]^{-1} = \frac{1}{\det(A)} \begin{bmatrix} d & -b \\ -c & a \end{bmatrix}, \tag{2.8}$$

where multiplying a matrix by a number consists of multiplying each entry by that number. It is now easy to verify that

$$[A][A]^{-1} = [A]^{-1}[A] = [I]. \tag{2.9}$$

This corresponds to the intuitive fact that if we apply a transformation and follow this by the inverse transformation we revert to the original configuration. We remark that, in general, the product of two different matrices is not commutative. In other words, the order of the application of two linear transformations matters.

Exercise 2.2 Verify that the product $[B][A]$ of the matrices in Example 2.1 is in general different from the product $[A][B]$ shown therein.

Exercise 2.3 Using the definition of Eq. (2.7) and the matrices of Example 2.1, verify that

$$\det(AB) = \det(BA) = \det(A)\,\det(B). \tag{2.10}$$

The *transpose* of a matrix $[A]$ is the matrix $[A]^T$ obtained by interchanging the rows with the columns of A. Thus, for example,

$$[A]^T = \begin{bmatrix} a & b \\ c & d \end{bmatrix}^T = \begin{bmatrix} a & c \\ b & d \end{bmatrix}. \tag{2.11}$$

Exercise 2.4 Show that $\det(A^T) = \det(A)$.

Exercise 2.5 Show that $([A][B])^T = [B]^T[A]^T$, and $([A][B])^{-1} = [B]^{-1}[A]^{-1}$.

Exercise 2.6 Verify that, with the operation of matrix multiplication and with the definition of unit and inverse matrix provided above, the collection of all non-singular square matrices of order 2 is a group. It is known as the *general linear group of order 2*, denoted by $GL(2, \mathbb{R})$.

Exercise 2.7 Show that a linear transformation (in two dimensions) represents a rigid rotation (of vectors) if, and only if, its matrix representation is of the form

$$[R] = \begin{bmatrix} \cos\theta & -\sin\theta \\ \sin\theta & \cos\theta \end{bmatrix}, \tag{2.12}$$

where sin and cos are the standard trigonometric functions. Verify that the collection of all matrices of this form constitutes a group. It is known as the *proper orthogonal subgroup* O_2 of $GL(2, \mathbb{R})$.

Exercise 2.8 Show that a square matrix of order 2 represents a rotation if, and only if, its inverse is equal to its transpose and if its determinant is equal to $+1$.

All the considerations above can be extended to three (or more) dimensions, except that the definitions of determinant and inverse are somewhat more complicated. The number of rows (or columns) of a square matrix is known as the *order* of the matrix. A square matrix of order n is said to be *singular* if its determinant vanishes. The collection of all non-singular square matrices of order n, with the operation of matrix multiplication, is a group known as the *general linear group of order* n, denoted by $GL(n, \mathbb{R})$. In the n-dimensional Euclidean space $\mathbb{R}^n$ a vector $\{v\}$ with components $v_1, v_2, \ldots, v_n$ is transformed by a matrix $[F] \in GL(n, \mathbb{R})$ into the vector $\{v'\}$ with components $v'_1, v'_2, \ldots, v'_n$ obtained by the matrix multiplication

$$\{v'\} = [F]\{v\} \tag{2.13}$$

or, more explicitly,

$$v'_i = \sum_{k=1}^{n} f_{ik} v_k, \qquad i = 1, \ldots, n. \tag{2.14}$$

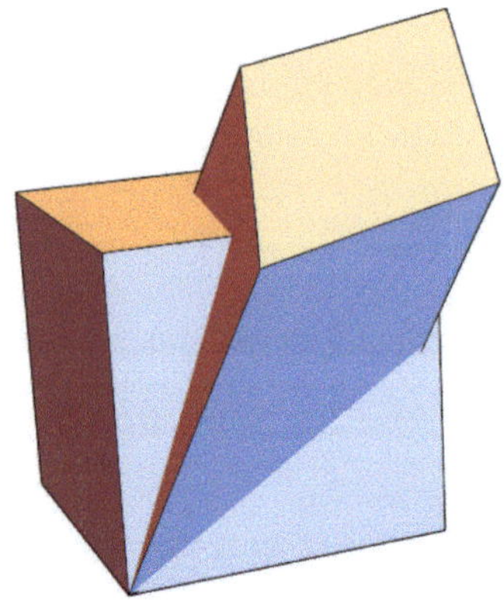

Fig. 2.4 Linear transformation of a unit cube

A three-dimensional counterpart of Fig. 2.3 is shown in Fig. 2.4. This particular linear transformation was obtained with the 3×3 matrix

$$[F] = \begin{bmatrix} f_{11} & f_{12} & f_{13} \\ f_{21} & f_{22} & f_{23} \\ f_{31} & f_{32} & f_{33} \end{bmatrix} = \begin{bmatrix} 0.8 & 0.3 & 0.3 \\ 0.2 & 0.9 & -0.4 \\ 0.1 & 0.4 & 1 \end{bmatrix}. \tag{2.15}$$

Just as in the case of two dimensions, the vanishing of the determinant of a linear transformation $[F]$ would indicate a dimensional collapse (namely, the cube would be squashed into a planar shape or even a line). The determinant of $[F]$ is given by *Cramer's rule* which results in

$$\begin{aligned} \det(F) = & f_{11}f_{22}f_{33} + f_{21}f_{32}f_{13} + f_{31}f_{12}f_{23} \\ & - f_{31}f_{22}f_{13} - f_{11}f_{32}f_{23} - f_{21}f_{12}f_{33}. \end{aligned} \tag{2.16}$$

The inverse is obtained as

$$[F]^{-1} = \frac{1}{\det(F)} \begin{bmatrix} f_{22}f_{33} - f_{23}f_{32} & f_{13}f_{32} - f_{12}f_{33} & f_{12}f_{23} - f_{13}f_{22} \\ f_{23}f_{31} - f_{21}f_{33} & f_{11}f_{33} - f_{13}f_{31} & f_{13}f_{21} - f_{11}f_{23} \\ f_{21}f_{32} - f_{22}f_{31} & f_{12}f_{31} - f_{11}f_{32} & f_{11}f_{22} - f_{12}f_{21} \end{bmatrix}. \tag{2.17}$$

2.2.1 *Determinants, Areas, Volumes

What is the physical meaning, if any, of the determinant of a linear transformation? Let us start from a two-dimensional case, such as that depicted in Fig. 2.3, in which a linear transformation was applied to the unit square to produce the parallelogram

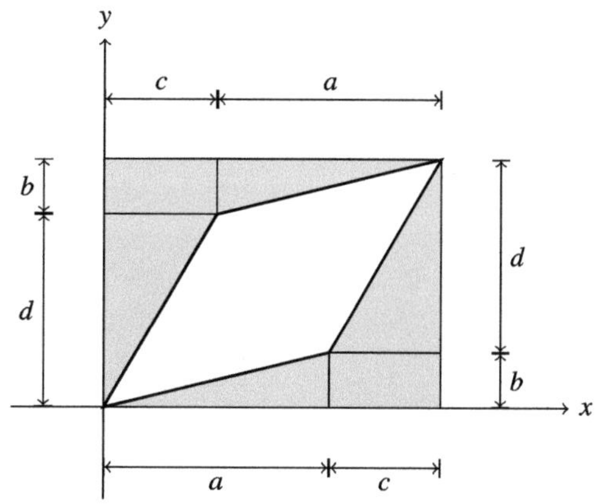

Fig. 2.5 Area of parallelogram

of Fig. 2.5. The area of the white parallelogram enclosed by the shaded areas is obtained as

$$(a + c)(b + d) - 2(bc + ab/2 + cd/2) = ad - bc, \tag{2.18}$$

which is exactly the determinant of the matrix A of Eq. (2.2). This is not a very convincing argument to conclude that there is a geometric meaning to the determinant, since it is only based on an example. We will, therefore, pursue a more careful line of thought. In the general case of an $n \times n$ matrix $[F]$, the determinant is defined as the sum of $n!$ terms. Each of these terms is a product of the form $f_{i_1,1} f_{i_2,2} \cdots f_{i_n,n}$, while the rows form a permutation $\{i_1, i_2, \ldots, i_n\}$ of $\{1, 2, \ldots, n\}$. Moreover, each term is multiplied by $+1$ or -1 according to the signature of this permutation. For $n = 3$ this policy is displayed explicitly in Eq. (2.16). A compact expression of the determinant is obtained introducing the *Levi–Civita* symbol defined as

$$\epsilon_{i_1,\, i_2\,, \ldots,\, i_n} = \begin{cases} +1 \text{ if } \{i_1,\ i_2\,,\ \ldots\ ,\ i_n\} \text{ is an even permutation} \\ -1 \text{ if } \{i_1,\ i_2\,,\ \ldots\ ,\ i_n\} \text{ is an odd permutation} \\ 0 \quad\ \text{otherwise} \end{cases}. \tag{2.19}$$

The determinant is obtained as

$$\det(F) = \sum_{i_1=1}^{n} \cdots \sum_{i_n=1}^{n} \epsilon_{i_1,\, i_2\,, \ldots,\, i_n}\ f_{i_1,1}\ \cdots\ f_{i_n,n}. \tag{2.20}$$

Exercise 2.9 Verify that Eq. (2.20) delivers the same results as Eq. (2.7) for $n = 2$, and Eq. (2.16) for $n = 3$.

Exercise 2.10 Prove that

$$\sum_{i_1=1}^{n} \cdots \sum_{i_n=1}^{n} \epsilon_{i_1,\, i_2\,,\, \dots\,,\, i_n}\, f_{k_1,i_1}\ \cdots\ f_{k_n,i_n} = \det(F)\ \epsilon_{k_1,\, k_2\,,\, \dots\,,\, k_n}. \tag{2.21}$$

Exercise 2.11 Verify that the determinant of a matrix is the same as the determinant of its transpose, and that the determinant of a product of matrices is equal to the product of their determinants.

Exercise 2.12 Using Eq. (2.20) show that the determinant of a matrix is not affected when a column is replaced by itself plus a linear combination of the other columns. [Hint: notice that the added terms produce results that are all symmetric in a pair of indices, whereas the Levi–Civita symbol is completely skew-symmetric.]

To interpret the result of Exercise 2.12 we can regard each column of $[F]$ as a vector in $\mathbb{R}^n$. If to a column (the first one, say) we add a linear combination of the other columns, the added term is a vector in the subspace spanned by these other vectors (second, third, etcetera). Since the linear combination is arbitrary, we conclude that any of the n given vectors can be replaced at no cost to the determinant by a vector perpendicular to all the remaining vectors.[1] In other words, the given vectors can be replaced by an orthogonal set.[2] This means that the new matrix $[F']$, when pre-multiplied by its transpose, is a diagonal matrix $[D] = [F']^T[F']$. The diagonal entries are nothing but the squares of the lengths of the modified vectors. It follows that the determinant of $[D]$ is the square of the volume of the parallelepiped subtended by the original vectors. Therefore, the absolute value of the determinant of $[F]$ equals that volume.

Exercise 2.13 Show that under the transformation (2.15) the unit sphere with equation $x^2 + y^2 + z^2 = 1$ in Cartesian coordinates gets distorted into the rugby-ball-shaped ellipsoid with equation

$$\begin{aligned} &(1.37128x - 0.232859y - 0.504528z)^2 \\ &+ (-0.310479x + 0.996119y + 0.491591z)^2 \\ &+ (-0.0129366x - 0.375162y + 0.853816z)^2 = 1. \end{aligned} \tag{2.22}$$

The surface of the ellipsoid is shown in Fig. 2.6. Calculate its volume.

[1] Historically, long before determinants were introduced, the Italian mathematician Bonaventura Cavalieri (1598–1647) had formulated the so-called method of the indivisibles. Similar ideas had been already advanced much earlier by Archimedes (287–212 BCE). A mundane description of Cavalieri's principle states that the volume of a stack of papers is not altered by the 'shear deformation' obtained by a relative sliding displacement between the leaves.

[2] This procedure, or a slight variant of it, is known as the Gramm–Schmidt orthogonalization process.

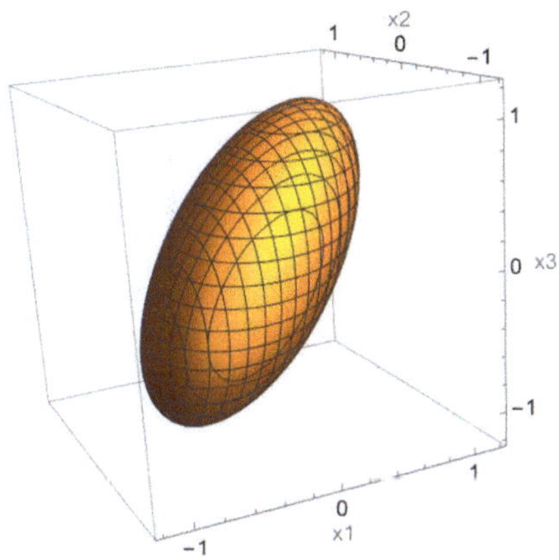

Fig. 2.6 Linear transformation of a unit sphere

2.2.2 *Some Subgroups of the General Linear Group*

The general linear group $GL(3, \mathbb{R})$ consists of all non-singular 3×3 matrices with real entries under the operation of matrix multiplication. Since matrices with vanishing determinant are excluded, the elements of this group fall naturally into one of two disjoint components according to the sign of their determinant.[3] The set of matrices with positive determinant is a subgroup denoted by $GL^+(3, \mathbb{R})$.

The subset of $GL(3, \mathbb{R})$ consisting of all matrices whose determinant has an absolute value of 1 is a subgroup sometimes referred to as the *unimodular group*. The *special linear group*, denoted by $SL^+(3, \mathbb{R})$, is the subgroup of $GL^+(3, \mathbb{R})$ consisting of all the matrices with determinant equal to $+1$. According to the interpretation given in the previous section, these matrices represent volume-preserving linear transformations. The distinction between the unimodular group and the special linear group is that in the latter not only the volume but also the handedness is preserved. In other words, the unimodular group includes transformations that reverse the orientation of a vector triad.

A matrix $[F]$ is said to be *orthogonal* if its inverse coincides with its transpose. This property immediately implies that its determinant is either $+1$ or -1. In dimension 3 orthogonal matrices with determinant $+1$ represent spatial rotations. These matrices constitute a group O_3 known as the proper orthogonal subgroup of $GL(3, \mathbb{R})$. If a rotation is preceded (or followed) by a reflection about a plane we obtain an orthogonal matrix with determinant -1. Orthogonal transformations are

[3] From the topological point of view the general linear group is not connected in the sense that it is not possible to go continuously from a matrix with positive determinant to one with negative determinant without hitting a singular matrix along the way.

those that preserve all lengths and angles in space, but the proper orthogonal ones also preserve orientation (a right-hand glove does not become a left-hand one).

Exercise 2.14 Prove that a square matrix is orthogonal if, and only if, its rows (equivalently, its columns) are mutually orthogonal unit vectors (in a Euclidean space).

Exercise 2.15 Prove that a 3×3 matrix is orthogonal if, and only if, it preserves the dot product of vectors in $\mathbb{R}^3$. Justify the conclusion that proper orthogonal matrices represent rigid-body rotations.

Exercise 2.16 The 2D rotation matrix of Exercise 2.7 can be reinterpreted in a right-handed Cartesian coordinate system as a 3D rotation (of an object) about the z-axis, that is,

$$[R]_z = \begin{bmatrix} \cos\theta & -\sin\theta & 0 \\ \sin\theta & \cos\theta & 0 \\ 0 & 0 & 1 \end{bmatrix}. \tag{2.23}$$

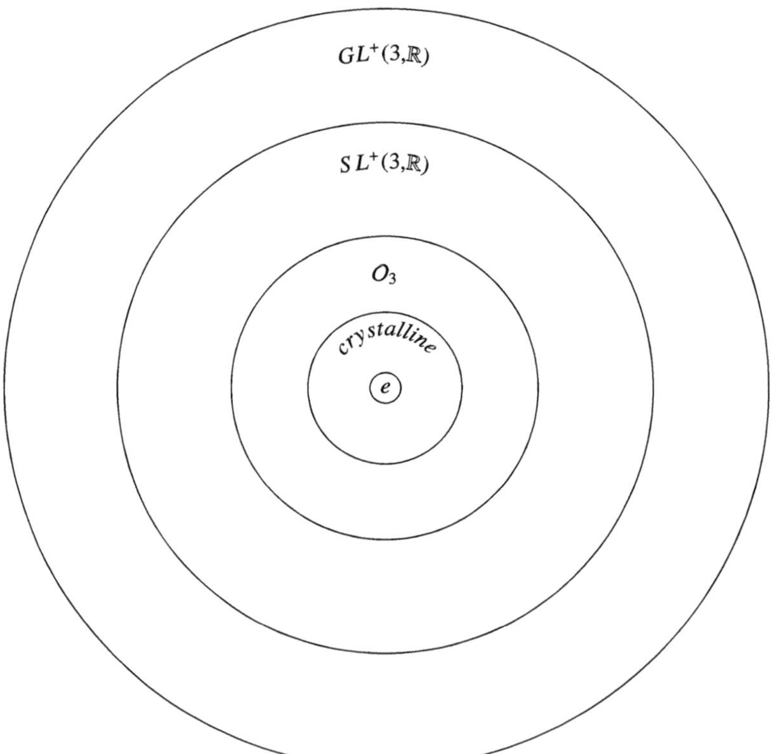

Fig. 2.7 Some subgroups of $GL^+(3, \mathbb{R})$

If the angle θ is positive, this rotation is counter-clockwise with respect to an observer standing along the positive direction of the z-axis. Obtain the 3×3 proper orthogonal matrices corresponding to counter-clockwise rotations about each of the coordinate axes. Verify that the result of applying a rotation of 90° counter-clockwise about z followed by the same amount of rotation about x is different from the result obtained by applying the rotations in the opposite order. Recall that the first transformation applied is the second factor in the product. You may want to verify this experimentally by rotating a book and following its spine. The conclusion of this exercise is that, while O_2 is commutative, O_3 is not. This is what makes rigid-body dynamics in space much more interesting than its planar counterpart.

The collection of all rotations about a fixed axis (just as in Exercise 2.16) constitutes a commutative subgroup of O_3. These subgroups are important in applications to layered materials, such as sedimentary rock, polar ice, and some composites. All the subgroups described so far in this section are continuous.[4] Other important subgroups are finite subgroups of O_3. They appear in the natural atomic arrangements of crystalline solids. A crystal with no symmetries beyond the identity $\{e\}$ is known as *triclinic*.

Figure 2.7 shows a schematic arrangement of some of the subgroups of $GL^+(3, \mathbb{R})$ defined so far. There are many other important subgroups not shown in this scheme, including all the possible conjugates of orthogonal subgroups, and other subgroups which include some orthogonal and some non-orthogonal elements.

Remark 2.1 It can be shown [28] that there is no intermediate subgroup between O_3 and $SL^+(3, \mathbb{R})$, that is, there is no group properly containing the former and properly contained in the latter.

Exercise 2.17 Show that $SL^+(3, \mathbb{R})$ and $\{e\}$ are not affected by conjugation. In other words, conjugates of these subgroups do not produce new subgroups.

Exercise 2.18 (*Inherited Symmetries) An ellipsoid $\mathcal{E}$ can be obtained as the image of the unit sphere $\mathcal{S}$ by applying a linear transformation $[F]$ to $\mathbb{R}^3$, as was done in Exercise 2.13. The sphere $\mathcal{S}$ enjoys all the symmetries of rotation about the origin, namely, its symmetry group is O_3. The image $\mathcal{E}$ of $\mathcal{S}$ by $[F]$, however, has only a finite group of rotational symmetries, namely, the identity and the 180° rotations about its three principal axes. Show that it has nevertheless inherited all the original symmetries in a conjugate sense. More precisely, every element of the conjugate group

$$O_{3,F} = [F]\, O_3\, [F]^{-1} \tag{2.24}$$

is a bijection of $\mathcal{E}$ onto itself. Generalize this important result.

---

[4] Technically, they are not only infinite (in terms of the number of group elements) but they are also differentiable manifolds. They are known as *Lie groups*.

Chapter 3
The Material Groupoid

Abstract After a cursory review of the concept of elasticity, while keeping in mind the motivation of the previous chapter, a definition of the material symmetry group at a point of an elastic body is formulated. For the sake of completeness, some basic notions of the mechanics of continuous media are reviewed and discussed, including a characterization of some elastic materials properties (such as fluidity, solidity, isotropy, orthotropy) according to the corresponding symmetry groups. The material symmetry group is a strictly local concept at each point of the body. The comparison of pairs of different points gives rise to the idea of the material groupoid. Two points made of the same material are said to be materially isomorphic. Material isomorphism can thus be regarded as a distant symmetry. A body all whose points are mutually materially isomorphic is said to be materially uniform.

3.1 Elastic Behaviour

3.1.1 *What Is Elasticity?*

In Chaps. 1 and 2 the general concepts of group and groupoid were introduced. How are these ideas applicable within the context of solid mechanics, beyond the skin-transplant imagery? We can visualize a very small piece of material (an infinitesimal cube, say) taken to a laboratory where appropriate equipment is available to apply any system of forces on this sample and to record the resulting deformations. Since the cubic sample is assumed to be very small its deformed shape can be regarded as a parallelepiped, just as depicted in Fig. 2.4. Accordingly, we may say that the deformation is completely characterized by a linear transformation. When represented in terms of components in a coordinate system x, y, z, therefore, we may say that a deformation of the sample is always described by a non-singular 3×3 matrix with positive determinant.

Many materials, such as rubber, steel, and wood, can be regarded as *elastic*, at least within a reasonable range of applied forces. A rubber ball subjected to the action of applied forces at its surface, will immediately return to its original

© The Author(s), under exclusive license to Springer Nature Singapore Pte Ltd. 2026

M. Epstein, *An Elementary Introduction to* n*-fold Groupoids*,
Surveys and Tutorials in the Applied Mathematical Sciences 19,
https://doi.org/10.1007/978-981-95-1869-2_3

configuration when the forces are removed. If subjected again to the same forces, it will attain again the same shape, regardless of the number of repetitions of the loading–unloading cycle. Nor would it matter whether the forces are applied rapidly or slowly. The result will always be the same.

The fact that the rubber ball returns immediately to its original shape may be interpreted in terms of a form of energy stored in the ball and completely recoverable as mechanical work, just like the case of a coil spring. This kind of energy stored in the deformed configuration of an elastic solid is known as *elastic energy*. From a thermodynamic viewpoint, we may say that the process of deformation of an elastic solid is a *reversible process*, involving no energy dissipation. If a coil spring is compressed and placed inside a rigid box, as soon as the spring is taken out of the box the elastic energy can be recovered and exploited to push a cart or lift a weight, regardless of the amount of time elapsed between the initial compression and the final release of the spring. This is not the case with other materials such as silly putty or plasticine.

We may, therefore, say that the *material response* of an elastic material is completely characterized by an *energy density function* $W : GL^+(3, \mathbb{R}) \to \mathbb{R}$

$$W = W(\mathbf{F}), \tag{3.1}$$

where, for simplicity, we have replaced the matrix symbol $[F]$ with the bold-face symbol $\mathbf{F}$. This function assigns a real number to each non-singular 3×3 matrix. The meaning of the result W is an energy per unit mass (so as to make this result independent of the size of the small sample used).[1] For a real body $\mathcal{B}$ of finite extent, the elastic energy density function depends also on the point, namely,

$$W = W(\mathbf{F}, X), \tag{3.2}$$

where X denotes the point $X \in \mathcal{B}$. Equation (3.2) is known as the *constitutive equation* of the elastic body $\mathcal{B}$.

3.1.2 Material Symmetries

Let us imagine that the function $W(\mathbf{F})$ in Eq. (3.1) is obtained experimentally by subjecting a very small material sample to linear deformations $\mathbf{F}$, while the resulting energy density $W(\mathbf{F})$ is measured. If the same experiment is performed on the same sample but preceded by a fixed linear deformation $\mathbf{G}$, so that the total deformation is $\mathbf{FG}$, the result obtained is obviously $W(\mathbf{FG})$, where W is the same function as before but evaluated at $\mathbf{FG}$ for each $\mathbf{F}$, while the pre-applied deformation $\mathbf{G}$ is maintained.[2]

[1] A more general notion of elasticity can be formulated in terms of the *stress tensor* rather than the scalar energy density. Materials for which an elastic energy density function exists are known as *hyperelastic*.

[2] For example, we may imagine that a small material sample is rotated by $\mathbf{G}$ before placing it back in the testing machine and repeating the experiment.

We say that $\mathbf{G}$ is a *material symmetry* if the condition

$$W(\mathbf{FG}) = W(\mathbf{F}) \tag{3.3}$$

is satisfied identically for all $\mathbf{F} \in GL^{+}(3;\mathbb{R})$, that is, for all possible linear deformations $\mathbf{F}$.

Exercise 3.1 Prove that the collection of all material symmetries of a given constitutive function $W(\mathbf{F}, X)$ constitutes a group $\mathcal{G}$. It is known as the *material symmetry group* at X.

3.1.3 Frame Indifference

A material symmetry, according to its very definition, is a fixed deformation $\mathbf{G}$ applied *before* submitting the sample to the variable deformation $\mathbf{F}$. It is for this reason that, following the standard convention for composition of functions, $\mathbf{G}$ appears to the right of $\mathbf{F}$ in the product $\mathbf{FG}$. In other words, a material symmetry is something one does in the material itself before subjecting it to further deformations $\mathbf{F}$. Are there also *spatial* symmetries? If so, they would represent an additional deformation that is applied *after* the variable deformation $\mathbf{F}$. This additional deformation would appear to the left of $\mathbf{F}$.

The *principle of material frame indifference*, sometimes also called the *principle of objectivity*, states that *all* constitutive equations are unaffected by the application of an arbitrary additional purely spatial rotation. Mathematically,

$$W(\mathbf{QF}, X) = W(\mathbf{F}, X), \qquad \forall \mathbf{Q} \in O_3. \tag{3.4}$$

The physical basis for this strong statement can be understood both in a passive or in an active way. In the first interpretation, we can consider two classical observers spatially rotated with respect to each other by the orthogonal matrix $\mathbf{Q}$. The principle states that these two observers should measure exactly the same energy density. The active interpretation, perhaps easier to understand, states that applying a subsequent rigid-body motion (rotation) to a deformed body does not alter any elongation or distortion, and thus the energy stored in the body remains unaffected. It has been argued, however, that if the material is made, at the microscopic scale, of molecules that include gyroscopic effects (which can certainly detect rigid-body motions) the principle would no longer be valid.

Exercise 3.2 Prove that if the dependence of W on $\mathbf{F}$ is of the form $W(\mathbf{F}^T\mathbf{F}, X)$, the principle of frame indifference is satisfied automatically. This condition can be shown to be also necessary. Question: would a function of the form $W(\mathbf{FF}^T, X)$ automatically satisfy the principle of frame indifference? Why, or why not? Can you think of another use for this last kind of dependence?

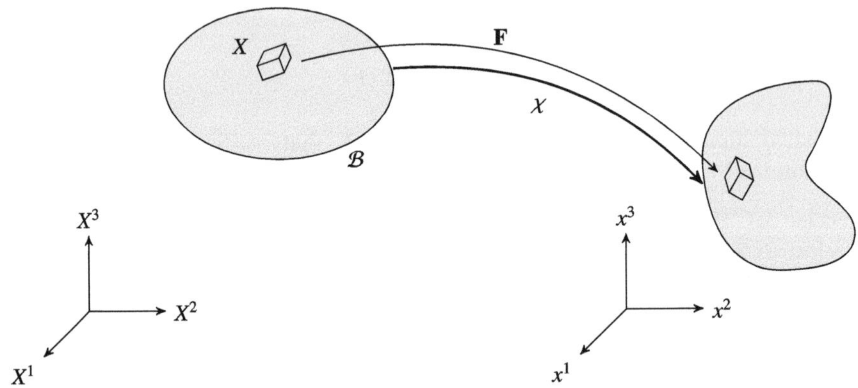

Fig. 3.1 A deformation χ, and its gradient $\mathbf{F}$ at point X

3.1.4 *Change of Reference Configuration

The constitutive function W in Eq. (3.2) contains an implicit dependence on the configuration adopted as reference. This subtle point is evidenced when we consider that a material body per se is nothing more than a continuous collection of material particles, what mathematicians would call a *topological manifold*.[3] This material entity makes itself amenable to observation and measurement only through its manifestations in our physical space. Each such manifestation is called a *configuration*.

For the purpose of defining and measuring quantities associated with a given configuration, it is usual to adopt one of these configurations as a *reference configuration* and to identify with it the body $\mathcal{B}$ itself. All other configurations can be referred to it via a *deformation*, which is a smooth injective map

$$\chi : \mathcal{B} \to \mathbb{R}^3. \tag{3.5}$$

Exercise 3.3 Discuss what would happen if the injectivity condition were not imposed on deformations.

If, as suggested in Fig. 3.1, the coordinates of a point X in the reference configuration $\mathcal{B}$ are denoted as X^I $(I = 1, 2, 3)$, while their counterparts in a deformed configuration $\chi(\mathcal{B})$ are x^i $(i = 1, 2, 3)$, the deformation χ in Eq. (3.5) boils down to three functions of three independent variables, namely,

$$x^i = x^i(X^1, X^2, X^3), \qquad i = 1, 2, 3. \tag{3.6}$$

[3] The stronger structure of a *differentiable manifold* is usually imposed. For the notion of configuration to make global sense, however, the existence of a global coordinate chart is also assumed.

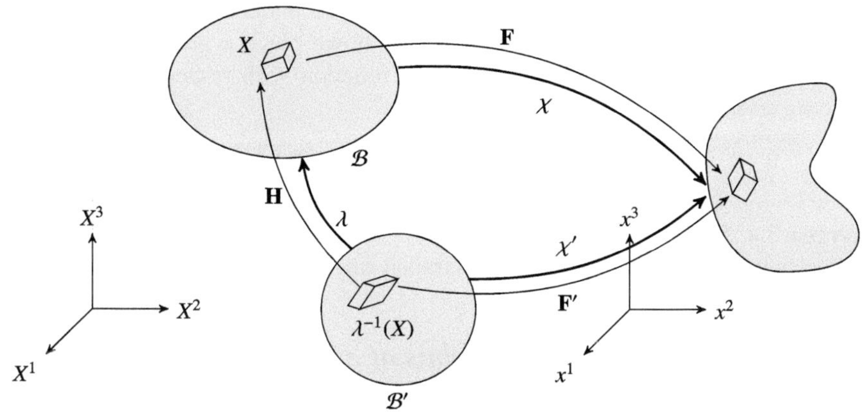

Fig. 3.2 Change of reference configuration

To describe a *motion*, a fourth independent variable, the time t, is added.

As far as an infinitesimal neighbourhood of a point $X \in \mathcal{B}$ is concerned, the deformation χ can be approximated by its *derivative map* at X. As we know from calculus of several variables this approximation is a non-singular linear transformation $\mathbf{F}$, called the *deformation gradient at* X, which is represented by a matrix whose entries are the partial derivatives of the three functions in Eq. (3.6) with respect to each of the independent variables. In other words, the entries F^i_J of this matrix (also known as the *Jacobian matrix* of the deformation) are

$$F^i_J = \frac{\partial x^i}{\partial X^J}, \qquad i, J = 1, 2, 3. \tag{3.7}$$

This equation connects what was said in Sect. 2.2 with the more general treatment. But what about the choice of reference configuration? How is the response function W affected by a different choice? Let two different reference configurations be given, denoted by $\mathcal{B}$ and $\mathcal{B}'$ in Fig. 3.2. The deformed configuration can be described via either one of the two deformations χ or χ'. But, since the first reference configuration can be regarded as a deformation λ of the second, we can obtain the relation between χ and χ' by a straightforward composition of maps, namely,

$$\chi' = \chi \circ \lambda. \tag{3.8}$$

As we know from calculus, the Jacobian of a composition of two maps is equal to the product of the Jacobians of the maps. In our case,

$$\mathbf{F}' = \mathbf{F}\,\mathbf{H}, \tag{3.9}$$

where $\mathbf{F}$ and $\mathbf{F}'$ are the deformation gradients with respect to the indicated configurations and $\mathbf{H}$ is the deformation gradient of λ.

How is the energy density function transformed under a change of reference configuration? The answer is now clear–cut. If we place a prime to distinguish between the expressions of the constitutive function with respect to each of the reference configurations adopted, we can write

$$W(\mathbf{F}, X) = W'(\mathbf{FH}, \lambda^{-1}(X)), \qquad \forall \mathbf{F} \in GL^{+}(3, \mathbb{R}). \tag{3.10}$$

Exercise 3.4 Show that under a change of reference configuration the material symmetry group at each point $X \in \mathcal{B}$ is transformed by conjugation.

3.1.5 *Constitutive Laws as Orbits of a Group Action

In our very first example of the concept of group, in Chap. 1, we had a chance to remark that a group (as an abstract concept) is often embodied in terms of its effect on a set. Thus every element of a certain group of rotations resulted in a *transformation* of the five-pointed star, namely, in a bijection of the star onto itself.

Definition 3.1 A group $\mathcal{G}$ *acts on the left* on a set A if a map $L_g : A \to A$ is assigned to each $g \in \mathcal{G}$ satisfying the following conditions:

1. Identity: The map L_e corresponding to the group identity e is the identity map of A, that is $L_e(x) = x$ for all $x \in A$.
2. Composition: The map assigned to the product gh of any two group elements is the composition of the maps assigned to each of the factors according to the rule $L_g \circ L_h = L_{gh}$.

Exercise 3.5 Show that each of the maps L_g is a bijection, that is, a one-to-one map of A onto A. Moreover, for each $g \in \mathcal{G}$, we have that $L_{g^{-1}} = (L_g)^{-1}$. [Hint: to prove injectivity, assume that applying L_g to two different points of A produces the same result. Apply to it $L_{g^{-1}}$.]

Definition 3.2 The *orbit* of a point $x \in A$ is the set

$$\mathcal{G}x = \{L_g(x) \ : \ g \in \mathcal{G}\}. \tag{3.11}$$

In other words, the orbit of x under the left action of the group is the subset $\mathcal{G}x \subset A$ obtained by applying to x the left action L_g as g ranges over the whole group $\mathcal{G}$.

The orbits of a left action of a group $\mathcal{G}$ on a set A are mutually disjoint (two orbits either coincide or do not have any common point). The relation of belonging to the same orbit is an equivalence relation. Consequently, the orbits constitute the corresponding equivalence classes. The set whose elements are these equivalence classes is known as the *quotient set* of A by $\mathcal{G}$.

Example 3.1 If our planet is idealized as the surface of a perfect sphere E, the group O_p of rotations about the polar axis acts on E on the left. The orbit $O_p x$ of a point $x \in E$ is precisely the geographical parallel through x.

A *right action* $R_g : A \to A$ of a group $\mathcal{G}$ on a set A is defined in the same way as a left action, except that the composition condition is replaced by

$$R_g \circ R_h = R_{hg}. \tag{3.12}$$

The orbits under a right action are defined accordingly and denoted as $x\mathcal{G}$.

Exercise 3.6 Every group $\mathcal{G}$ has several canonical (or natural) actions on itself. Show that assigning to each $g \in \mathcal{G}$ the map $L_g : \mathcal{G} \to \mathcal{G}$ defined by

$$L_g(h) = g\, h\, g^{-1}, \tag{3.13}$$

we obtain a left action of $\mathcal{G}$ on itself. What are the orbits of this action? [Hint: refer to Eq. (1.16).]

Considering a constitutive equation of the form (3.1), the dependence of W on the independent variable $\mathbf{F}$ is typically obtained either experimentally or by virtue of some assumed model of ideal material response. In either case, the function W is presumed to belong to a space of functions $\mathcal{F}$, such as the set of all continuous real-valued functions on the general linear group $GL(3, \mathbb{R})$, that is, continuous functions $\phi(\mathbf{F})$ that assign to each non-singular matrix $\mathbf{F}$ a real number. We can now define the *right action* of $GL^+(3, \mathbb{R})$ on $\mathcal{F}$ as

$$R_H(\phi(\mathbf{F})) = \phi(\mathbf{FH}). \tag{3.14}$$

Exercise 3.7 Verify that the prescription (3.14) is indeed a right action on the set $\mathcal{F}$.

Comparing Eq. (3.14) with (3.10), we realize that the right action of the general linear group on $\mathcal{F}$ corresponds exactly to the modification of a constitutive equation brought about (at a point X) by a change of reference configuration. This kind of modification does not entail any physical change, but only a change of reference. We may say now that an elastic material point is characterized not just by a constitutive function W but rather by a whole orbit $W\,GL^+(3, \mathbb{R})$ of the right action of the general linear group. From a practical standpoint, this conclusion means that two points are made of the same elastic material if, and only if, their constitutive functions belong to the same orbit of this action. In Sect. 3.2 we will arrive at an equivalent conclusion after introducing the more intuitive notion of *material isomorphism* [29, 32] without the benefit of the discussion carried out in the present, somewhat more abstract, section.

3.1.6 Classification of Elastic Materials

On physical grounds, it is generally accepted that all material symmetries are volume-preserving. In other words, it is assumed that every material offers resistance to a volume change. What is the consequence of this assumption? In Sect. 2.2 we learned that the absolute value of the determinant of a linear transformation conveys the information about the relative change of volume between the deformed and the original element. Accordingly, we conclude that *the material symmetry group of a material point is necessarily a subgroup of the special linear group* $SL^+(3, \mathbb{R})$.

Remark 3.1 It is also acceptable to include symmetries that do not preserve orientation, in which case we can replace the special linear group with the whole unimodular group.

Exercise 3.8 At the risk of repeating the conclusions drawn from previous exercises, verify that the change brought about on the material symmetry group of a body point by a change of reference configuration does not lead to a violation of the principle that the determinant of all material symmetries remains equal to 1.

Isotropy

A material point is said to be *isotropic* if in some reference configuration[4] its material symmetry group contains the proper orthogonal group O_3. Such privileged configuration, if it exists, is known as an *undistorted state*. The name *isotropy* is meant to indicate that in an undistorted state it is impossible to distinguish by mechanical means whether the material has been rotated arbitrarily before performing a test. In short, the material in an undistorted state has no preferred directions [30].

Exercise 3.9 Go back to Exercise 3.2 and address the last question in that exercise.

According to Remark 2.1, there are only two kinds of isotropic elastic materials. The first kind is the *elastic fluid*. Its material symmetry group in any reference configuration is the special linear group $SL^+(3, \mathbb{R})$. To connect this mathematical statement with intuition, the fact that only preservation of volume is required can be interpreted as indifference to a change in shape. Only a change of volume affects the value of the elastic energy. As we were told in elementary school, a fluid obediently adopts the shape of its container.[5]

The second kind of isotropic material is the isotropic solid. To define an elastic solid, one may think of a material sample in a completely unloaded configuration.

---

[4] As far as a point is concerned, a change of reference configuration entails just a linear transformation. For this reason it is more appropriate to talk about a *local* reference configuration at X, which can be more rigorously defined as an equivalence class of configurations with a common gradient at X.

[5] In his acute observations on fluids, Leonardo da Vinci was fascinated by the fact that water can undergo changes of shape, without changing its volume, at no energetic cost. As better expressed by Walter Isaacson [21], "flowing water is constantly going through perfect geometric transformations. No wonder he loved it."

Not all materials may sustain such a configuration. A gas, for instance, does behave elastically but cannot sustain an unloaded configuration, since it requires the pressure applied by the walls of a container. The collection of all symmetries of this unloaded sample is a group $\mathcal{G}$. The elastic sample is said to be *solid* if its symmetry group is a subgroup of the orthogonal group $\mathcal{O}_3$. Physically this means that any non-rotational transformation of the sample will definitely affect its elastic energy content. The special case in which $\mathcal{G} = \mathcal{O}_3$ corresponds to an *isotropic solid*: its material response is unaffected by the pre-application of *any* rotation of the unloaded sample. This is, of course, a continuous group.

Remark 3.2 Since the local reference configuration adopted may not be an undistorted state, we conclude that a material is a solid if, and only if, its material symmetry group (in any configuration) is a conjugate (by any fixed element of the general linear group) of a subgroup of $\mathcal{O}_3$. A solid is isotropic if its material symmetry group (in any configuration) is a conjugate of $\mathcal{O}_3$. See Exercise 3.4 (or Exercise 3.12).

Aeolotropy

A material that is not isotropic is called *aeolotropic* or *anisotropic*. A common occurrence of solid aeolotropy is known as *transverse isotropy*. In an undistorted state a transversely isotropic solid has the symmetry group of all rotations about a fixed axis. This is a continuous commutative group. It can be shown that all the remaining subgroups of $\mathcal{O}_3$ are discrete. They comprise the crystal classes that manifest themselves macroscopically in properties such as *orthotropy*. A material with no symmetries except the trivial identity is said to be *triclinic*.

Exercise 3.10 (Orthotropy) In an orthotropic solid there are three mutually orthogonal preferred directions x, y, z. The corresponding symmetry group $\mathcal{R}^+$ consists of four elements: the identity and the 180°-rotations $[\hat{R}]_x$, $[\hat{R}]_y$, $[\hat{R}]_z$ about each of those axes. Show that all four elements can be obtained from just two of these rotations by compositions. We say that the finite group $\mathcal{R}^+$ is *generated* by the subset $\{[\hat{R}]_x, [\hat{R}]]_y\}$.

Remark 3.3 We have defined a solid via the proper orthogonal group $\mathcal{O}_3$. It is also possible to define it invoking the subgroup of all orthogonal matrices, regardless of whether their determinant is $+1$ or -1, namely including not only symmetries of rotation but also those of reflection.

Exercise 3.11 Orthotropy is often defined so as to include reflections about the coordinate planes. The corresponding group $\mathcal{R}$ consists of eight elements: the unit $[I]$, the inversion $[-I] = -[I]$, the three rotations $[\hat{R}]_x$, $[\hat{R}]_y$, $[\hat{R}]_z$ and the three reflections $[S]_x$, $[S]_y$, $[S]_z$ about the planes yz, xz and xy, respectively. Show that this group is generated by (a) the inversion and two reflections, or (b) the inversion and two rotations.

What else is left? We have described so far elastic fluids and solids. An elastic material that is neither fluid nor solid cannot be isotropic [28]. Wang [31] and Coleman [7] have studied hypothetical material models that may be called *fluid crystals* which share some properties with solids, but are also able to flow like fluids in certain preferred directions.

3.2 Material Isomorphisms and the Material Groupoid

Consider two points, X_1 and X_2 of $\mathcal{B}$. Are they made of the same material? To answer this question fairly, we have to consider the possibility that point X_2 may be replaced with a graft obtained by cutting a small piece (a cube say) around point X_1 and transferring it to point X_2 via a linear transformation $\mathbf{P}_{12}$, as shown in Fig. 3.3. Let the original material at X_2 be subjected to some linear transformation $\mathbf{F}$. This would result in an energy density given by $W(\mathbf{F}, X_2)$. But the piece transplanted from X_1 via $\mathbf{P}_{12}$ would experience a total deformation of $\mathbf{F}\mathbf{P}_{12}$. We conclude that the material at X_2 is the same as the material at X_1 if, and only if, there exists a transplant $\mathbf{P}_{12}$ such that the following condition is satisfied identically for all possible deformations $\mathbf{F}$:

$$W(\mathbf{F}, X_2) = W(\mathbf{F}\mathbf{P}_{12}, X_1). \tag{3.15}$$

Indeed, this equation ensures that one would be unable to distinguish between the material responses at both points. If this is the case we say that points X_1 and X_2 are *materially isomorphic*. The transplant $\mathbf{P}_{12}$ is called a *material isomorphism* between X_1 and X_2.

Given an energy density function $W(\mathbf{F}, X)$ for a body $\mathcal{B}$ it may happen that all pairs of points in the body are materially isomorphic. In that case we say that the body is *materially uniform* (or *uniform*, for short). Many elastic bodies found in applications (such as the so-called *functionally graded bodies*) are not materially uniform.

If we mentally draw an arrow for each material isomorphism between every ordered pair of body points, we obtain the *material groupoid* $\mathcal{Z} \rightrightarrows \mathcal{B}$ associated

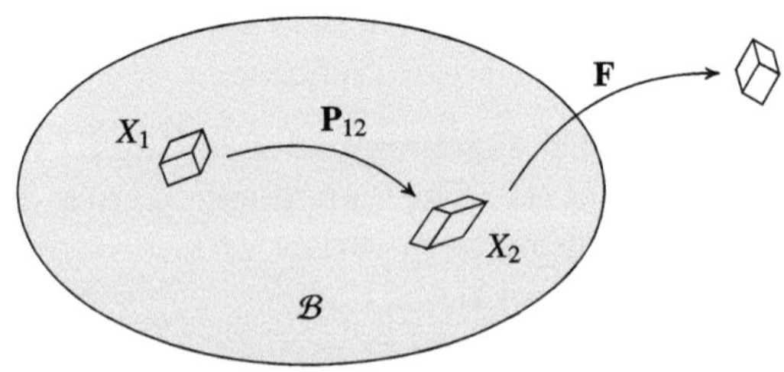

Fig. 3.3 A material isomorphism: (3.15) satisfied for all **F**

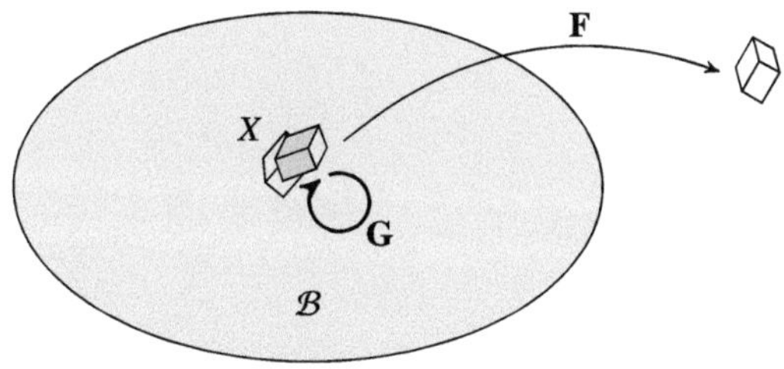

Fig. 3.4 A material symmetry: (3.16) satisfied for all **F**

with the energy function W. The body $\mathcal{B}$ is materially uniform if, and only if, its material groupoid is transitive, in accordance with Definition 1.4.

The vertex groups of the material groupoid represent local *material symmetries*. Indeed, if in Eq. (3.15) we identified $X_2 = X_1 = X$, we obtain

$$W(\mathbf{F}, X) = W(\mathbf{FG}, X) \tag{3.16}$$

for all deformations $\mathbf{F} \in GL(3, \mathbb{R})$, as suggested in Fig. 3.4. We may, therefore, think of a material symmetry as a linear self-implant $\mathbf{G}$ at $X \in \mathcal{B}$ that, when applied to a sample before performing an experiment to measure the energy density function, does not alter the result of the experiment regardless of the subsequent deformation $\mathbf{F}$ applied.

Exercise 3.12 Using just Eqs. (3.15) and (3.16), prove that the material symmetry groups of two materially isomorphic points are conjugate.

Given a body $\mathcal{B}$, considered just as a continuum regardless of its material response, one can always define a groupoid $\Pi^1(\mathcal{B}, \mathcal{B}) \rightrightarrows \mathcal{B}$ whose arrows are all the possible transplants between all pairs of points. Thus, the collection of arrows from any point of $\mathcal{B}$ to any other is the general linear group $GL(3, \mathbb{R})$ or, equivalently, the collection of all non-singular 3×3 matrices. This groupoid is known as the *1-jet groupoid over* $\mathcal{B}$. Clearly, every material groupoid is a subgroupoid of $\Pi^1(\mathcal{B}, \mathcal{B}) \rightrightarrows \mathcal{B}$.

3.2.1 The Material Groupoid of a Uniform Body

If the body happens to be uniform, the material groupoid is transitive and the symmetry groups at all points are mutually conjugate. In this case, it is convenient to think (just as in the case of the skin graft and the donor source) of an *archetypal material element* placed outside the body, as shown in Fig. 3.5. Since all the vertex groups are mutually conjugate, it seems appropriate to say that the symmetry group $\bar{\mathcal{G}}$ of the archetype can be considered as the *archetypal symmetry group* of the uniform body. Upon implanting the archetype into each point $X \in \mathcal{B}$ via a field

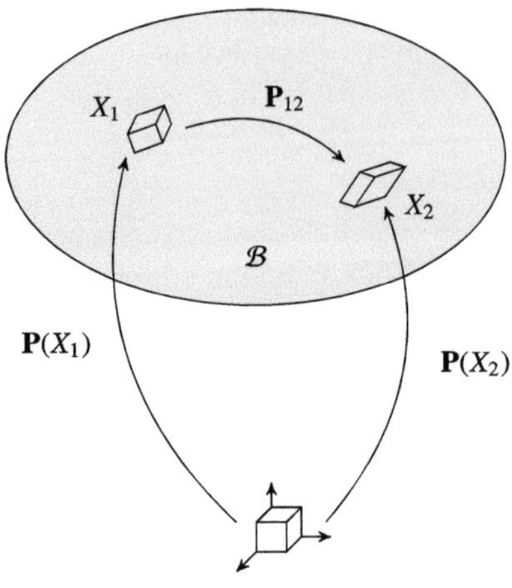

Fig. 3.5 An archetype and its implants

of *implant maps* $\mathbf{P}(X)$, the transplant map $\mathbf{P}_{12}$ between X_1 and X_2 is recovered as the composition

$$\mathbf{P}_{12} = \mathbf{P}(X_2)\ (\mathbf{P}(X_1))^{-1}\,. \tag{3.17}$$

The symmetry group $\mathcal{G}_X$ at $X \in \mathcal{B}$ is obtained by conjugation as

$$\mathcal{G}_X = \mathbf{P}(X)\ \bar{\mathcal{G}}\ (\mathbf{P}(X))^{-1}. \tag{3.18}$$

If the archetype's constitutive equation is $\bar{W}(\mathbf{H})$, the constitutive equation of the body is

$$W(\mathbf{F}, X) = \bar{W}(\mathbf{FP}(X)). \tag{3.19}$$

Exercise 3.13 Prove Eq. (3.18) directly from Eq. (3.19). [Hint: if $\bar{\mathbf{G}}$ is a symmetry of $\bar{W}$ it follows that $\bar{W}(\mathbf{FP}(X)) = \bar{W}(\mathbf{FP}\bar{\mathbf{G}}(X))$.]

Remark 3.4 We have established that a body is materially uniform if, and only if, its material groupoid is transitive. We have concluded that in a uniform body all the vertex groups are mutually conjugate and that, therefore, we can speak of a typical (archetypal) symmetry group. The converse, however, is not true. In other words, conjugacy of the vertex groups does not imply uniformity. In physical terms, a body may be made of points that enjoy the same material symmetries (isotropy, say) without being uniform. An important example of this kind is provided by the so-called *functionally graded materials*, of great importance in many technological applications. The properties (such as the elastic moduli) vary continuously from point to point but may preserve the material symmetries. An extended concept of the material groupoid for functionally graded bodies has been presented in [12].

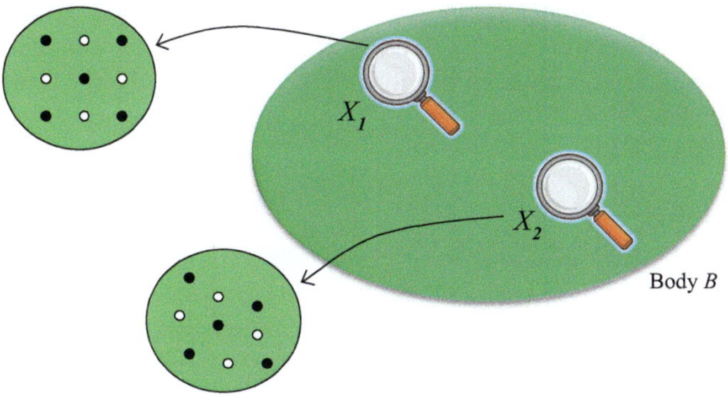

Fig. 3.6 Table salt

3.2.2 *Uniformity and Homogeneity

The physical meaning of material uniformity of a body $\mathcal{B}$ is that all points of $\mathcal{B}$ are made of *the same material*. An intuitive way to explain this concept without any mathematics is to imagine that the material is a solid crystal, say sodium chloride (or table salt). If we were able to observe its atomic structure under a powerful microscope, we would discern the expected regular arrangement of sodium and chlorine atoms, as sketched in Fig. 3.6. At two points, say X_1 and X_2, the alternation of the atoms of both elements takes place, thus guaranteeing the sameness of the chemical constitution at both places. Nevertheless, this topological identity is not matched by a geometric counterpart. Indeed, the distances between the atoms as well as the angles formed by the lattice segments are completely different in both neighbourhoods. If we were to transplant X_1 into X_2 we would need to effect some stretching and shearing. This is again a manifestation of the concept of transplantation in a somewhat different guise.

Having established that the body is uniform, however, there is still another unanswered question, namely: Can a smooth deformation of the body lead to a simultaneous straightening of the neighbourhoods of all the points of $\mathcal{B}$ such that they all look exactly like X_1? In other words, using the language of crystallography, can the atomic arrangement over the whole body be brought to a Bravais lattice configuration? A positive answer to this question is tantamount to the assertion that the body is *defect free* or *homogeneous*. A negative answer would mean that the body is *inhomogeneous*, that is, it contains defects.

Isolated defects are the subject of the theory of crystal dislocations. When the number of defects per unit volume is very large, such as in the study of metal plasticity, it is possible to talk about *continuous distributions of dislocations*. Differential geometric tools can be used to establish faithful measures of dislocation densities as well as criteria for the body to be homogeneous [13, 29, 32]. We will not pursue this topic any further, but the solution of Exercise 3.15 below can be used as a motivating introduction to the theory.

Example 3.2 (Inhomogeneity) A stress-free archetype of a solid can be regarded as the orthonormal natural basis $\mathbf{E}_\alpha$ ($\alpha = 1, 2, 3$) of $\mathbb{R}^3$. A field of implants $\mathbf{P}(X)$ for a uniform body, such as depicted in Fig. 3.5, is, therefore, a field of bases $\mathbf{P}_\alpha$ ($\alpha = 1, 2, 3$) which can be represented in Cartesian components as

$$\mathbf{P}_\alpha = P_\alpha^I \, \mathbf{E}_I. \tag{3.20}$$

In this equation we are adhering to the *summation convention* according to which a diagonally repeated index is summed upon the range $I = 1, 2, 3$. These bases are not necessarily orthogonal nor unitary, so that the implants may involve not just rotations but also stretches and distortions. Only when the matrix $[P_\alpha^I]$ belongs to O_3 we have a case of a purely rotational implant. For the sake of a concrete example, let us consider a uniform body made of a material with a discrete symmetry group (such as orthotropy). If the implant field consists of pure rotations, it is clear that the body in the reference configuration is stress-free. If, in addition, this field is constant (that is, independent of X) we obtain a homogeneous configuration. We can visualize this case by thinking of the archetype as an infinitesimal Lego block repeatedly inserted in the body in parallel arrays, just like in a well-built brick wall. If, on the other hand, the rotation field is not constant, the body is still stress-free, but the blocks leave small gaps in between. In the infinitesimal limit these gaps become a continuous distribution of dislocations. The body is no longer homogeneous. For arbitrary (not necessarily purely rotational) fields of implants of the archetype, it is not difficult to derive differential-geometric criteria of *integrability* that would decide whether the body is homogeneous or not. These criteria check the possibility of deforming the body in a smooth way so that it can be brought to a perfect brick-wall configuration.

Exercise 3.14 Prove that in the case of an arbitrary rotation field (which for a discrete symmetry group would in general lead to the presence of dislocations), if the archetype is isotropic the body is homogeneous. [Hint: take advantage of the degree of freedom of rotation afforded by the group in the choice of the archetype implants.]

Exercise 3.15 Consider a two-dimensional uniform body with a triclinic archetype in a natural stress-free state. If the field of implants is purely rotational, namely if (in a referential Cartesian coordinate system X, Y)

$$[P] = \begin{bmatrix} \cos\theta(X,Y) & -\sin\theta(X,Y) \\ \sin\theta(X,Y) & \cos\theta(X,Y) \end{bmatrix}, \tag{3.21}$$

where $\theta(X, Y)$ is a differentiable function measuring the angle of rotation of the implanted frame, prove that the body is homogeneous if, and only if, $\theta(X, Y)$ is the constant function. Explain this result intuitively. Conclude that if the body were in the shape of an annular sector, and if we were to align the orthonormal bases with the polar coordinate lines, the body would be materially inhomogeneous. If, however, the implants were to consist of a combination of a rotation with a preceding

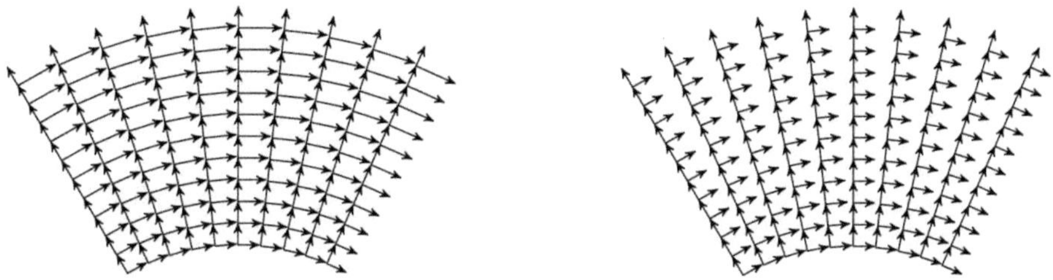

Fig. 3.7 Homogeneous (left), and inhomogeneous (right) body

elongation (or contraction) along the archetype's x direction so that the implanted frames would coincide with the *natural coordinate basis* of the polar coordinate system, the body would be homogeneous, as shown in Fig. 3.7.

Chapter 4
Double Groupoids

Abstract The set of objects of a groupoid may happen to be also a groupoid. A double groupoid is an even more involved entity. It consists of a set $\mathcal{Q}$ with two groupoid structures, each of which has a different groupoid, say $\mathcal{H}$ and $\mathcal{V}$, as its set of objects. These two base groupoids, however, have a common base $\mathcal{B}$. Accordingly, a double groupoid involves 4 sets intimately interconnected by 4 different groupoid structures. The arrows of a double groupoid $\mathcal{Q}$ may be regarded informally as some kind of hyper-arrows, each of which can be visualized as a square consisting of two arrows of $\mathcal{H}$ joining a pair of arrows of $\mathcal{V}$, or vice versa.

4.1 The Concept

There is no restriction on the nature of the set of objects (or base) of a groupoid. For this reason, it is almost inevitable to ask the question: what is a groupoid $\mathcal{Q}$ whose set of objects $\mathcal{H}$ is also a groupoid? What is an arrow in $\mathcal{Q} \rightrightarrows (\mathcal{H} \rightrightarrows \mathcal{B})$? Since the objects in $\mathcal{H}$ are themselves arrows, we may think of an arrow of $\mathcal{Q}$ as some kind of hyper-arrow connecting between two arrows of $\mathcal{H}$.

The notion of *double groupoid* crystallizes when those supposed hyper-arrows are identified with pairs of arrows of another groupoid $\mathcal{V} \rightrightarrows \mathcal{B}$ having the same set of objects $\mathcal{B}$ as $\mathcal{H}$. Moreover, one requires that the first arrow of the pair must connect the tails of the two arrows of $\mathcal{H}$, while the second arrow connects their tips. We think of an ideal *cell* (playing a role similar to that of the ideal arrow $\mathcal{S}$ described in Remark 1.1) which can be imagined as a square $\mathcal{S}_2$ whose sides are oriented unit segments in the plane with coordinates x_1, x_2, as shown in Fig. 4.1. The arrows along x_1 (horizontal arrows) are to be realized in the *horizontal side groupoid* $\mathcal{H}$, while those along x_2 (vertical arrows) will be realized as arrows in the *vertical side groupoid* $\mathcal{V}$.

The punchline comes upon ascertaining that the roles of $\mathcal{H}$ and $\mathcal{V}$ could be reversed, with a pair of arrows of $\mathcal{H}$ serving as hyper-arrows for a pair of arrows of $\mathcal{V}$. In more technical terms, what we have obtained is that the set $\mathcal{Q}$ has *two different groupoid structures*, one with base $\mathcal{H}$ and another one with base $\mathcal{V}$. These two

© The Author(s), under exclusive license to Springer Nature Singapore Pte Ltd. 2026
M. Epstein, *An Elementary Introduction to* n*-fold Groupoids*,
Surveys and Tutorials in the Applied Mathematical Sciences 19,
https://doi.org/10.1007/978-981-95-1869-2_4

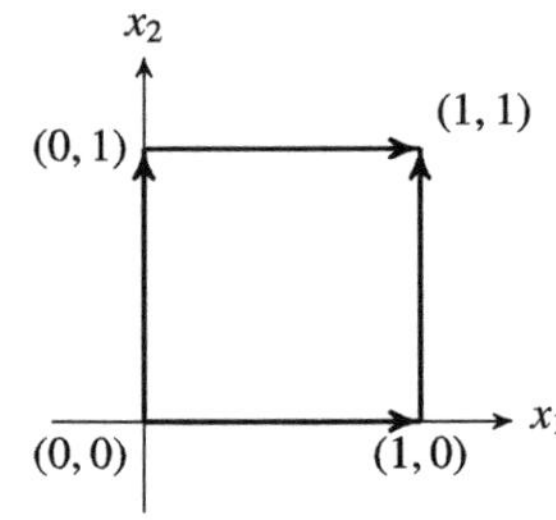

Fig. 4.1 The standard square cell $\mathcal{S}_2$

structures will be called, respectively, vertical and horizontal. The respective source and target projection maps will be indicated with a superposed tilde and the subscript V or H. The elements of Q will be called *squares* and denoted with double-strike letters ($\mathbb{A}, \mathbb{B}, \ldots$). Thus, for example, $\tilde{\alpha}_H(\mathbb{A}) \in \mathcal{V}$, and $\tilde{\alpha}_V(\mathbb{A}) \in \mathcal{H}$. In other words, the source of a horizontal (vertical) arrow pair is a vertical (horizontal) arrow. The diagram (4.1) is a brief representation of the structure of a double groupoid.

$$\begin{array}{ccc} Q & \underset{\tilde{\beta}_H}{\overset{\tilde{\alpha}_H}{\rightrightarrows}} & \mathcal{V} \\ \tilde{\alpha}_V \downdownarrows \tilde{\beta}_V & & \alpha_V \downdownarrows \beta_V \\ \mathcal{H} & \underset{\beta_H}{\overset{\alpha_H}{\rightrightarrows}} & \mathcal{B} \end{array} \tag{4.1}$$

Figure 4.2 depicts most of the concepts elaborated so far. The numbering of the corners of the idealized cell S_2 is naturally generated by interpreting the pairs of coordinates as a binary number. Thus, the corners with coordinates $(0, 0)$, $(0, 1)$, $(1, 0)$, $(1, 1)$ in Fig. 4.1 correspond, respectively, to the corners numbered as 0, 1, 2, 3 in Fig. 4.2. These corners get to be projected (via the composition of two projections) into four (not necessarily distinct) points W_0, W_1, W_2, W_3 of $\mathcal{B}$.[1]

Exercise 4.1 Verify that each of the four-sided diagrams made of dashed arrows in Fig. 4.2 is commutative. For example,

$$\alpha_V \circ \tilde{\beta}_H(\mathbb{A}) = \beta_H \circ \tilde{\alpha}_V(\mathbb{A}) = W_2. \tag{4.2}$$

Make sure that each projection has a definite meaning in the appropriate groupoid.

[1] Recall that the arrows in a groupoid are not in any sense 'directly above' the points on which they project. Arrows and points dwell in different worlds. They are only related by the projection maps (which do not have in general a Euclidean geometric meaning). The figures are meant only to help develop an intuitive feeling of the structures involved.

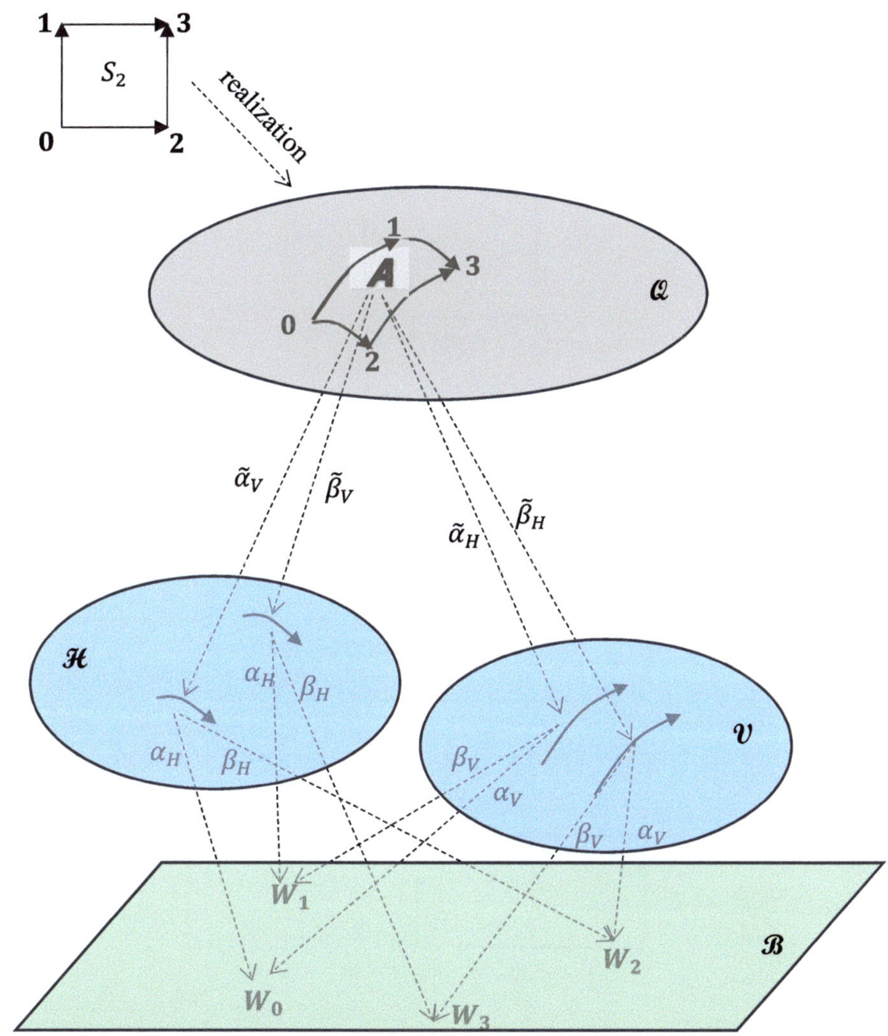

Fig. 4.2 Participants in a double groupoid

4.2 The Operations

To complete the description of the horizontal and vertical groupoid structures, it is necessary to define the respective composition laws, units, and inverses. Let $\mathbb{A}$ be a square in Q, as represented in (4.3). Its two horizontal sides s and t are obtained as source and target using the projections of the vertical structure, that is, with the maps $\tilde{\alpha}_V$ and $\tilde{\beta}_V$, respectively. Similarly, the vertical sides $\hat{s}$ and $\hat{t}$ are the source and target projections of $\mathbb{A}$ according to the horizontal-structure projections $\tilde{\alpha}_H$ and $\tilde{\beta}_H$.

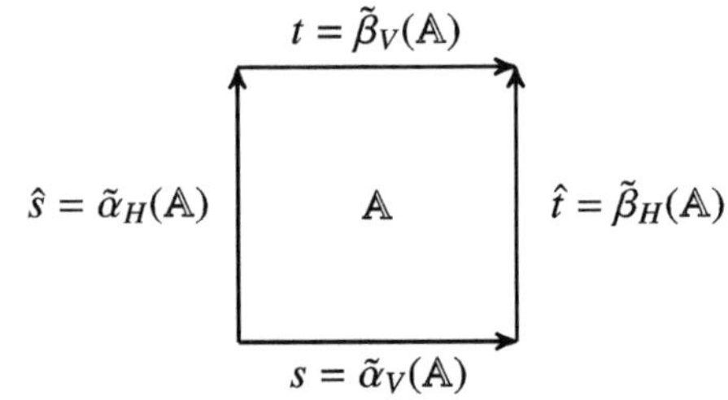

(4.3)

Since Q is endowed with two different groupoid structures, each structure must have its own well-defined product, which will be denoted by $\odot_H$ and $\odot_V$, respectively, for the horizontal and vertical structures. In the horizontal structure, the product of two squares, $\mathbb{A} \odot_H \mathbb{A}'$, can only be carried out if $\hat{s} = \hat{t}'$, in agreement with the usual concatenation criterion in a groupoid.

Exercise 4.2 Clarify what is meant by the last statement about 'the usual concatenation criterion in a groupoid'. What are the 'arrows' of the horizontal structure? What does the 'tip-to-tail' condition mean in this groupoid?

The composition law of the horizontal structure of Q is obtained by applying the composition law of the horizontal side groupoid $\mathcal{H}$ to each pair of adjacent horizontal arrows of the two squares. This is easily represented graphically as

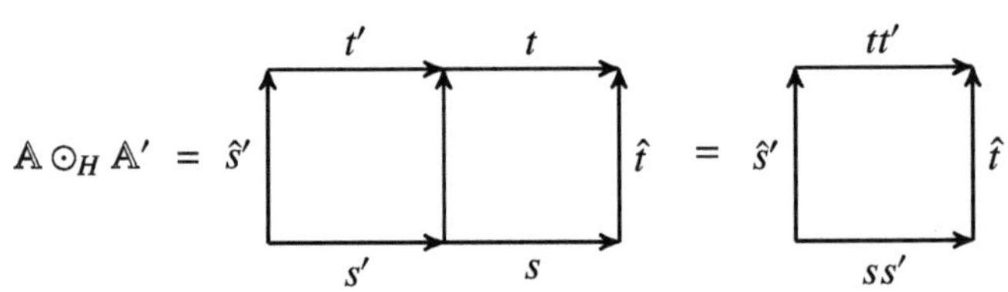

(4.4)

The unit square of the horizontal structure at the vertical arrow $\hat{u}$ is the square

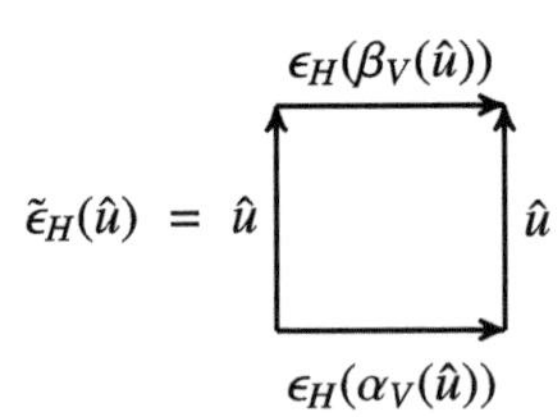

(4.5)

Exercise 4.3 Check that the unit square (4.5) indeed satisfies $\mathbb{A} \odot_H \tilde{\varepsilon}_H(\hat{u}) = \mathbb{A}$, where $\hat{u} = \tilde{\alpha}_H(\mathbb{A})$.

Similarly, in the vertical structure the product $\mathbb{A} \odot_V \mathbb{A}'$ can only be carried out if $s = t'$, or graphically,

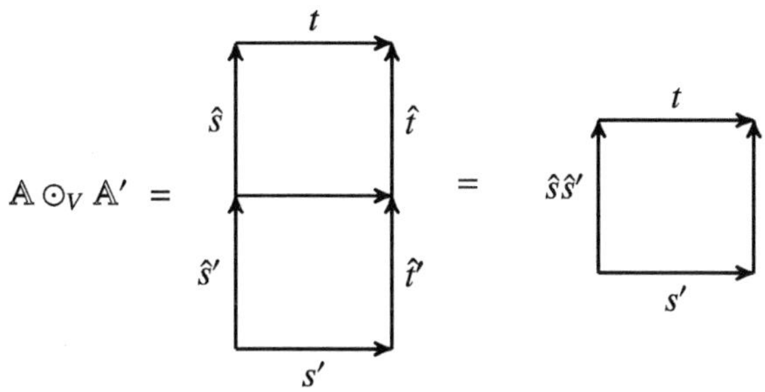

(4.6)

The unit square of the vertical structure at the horizontal arrow u is the square

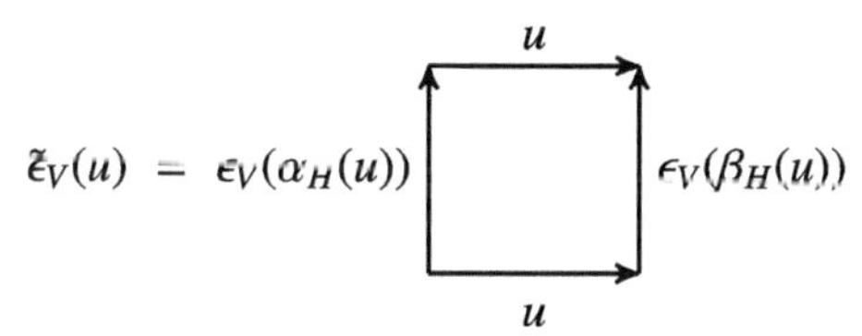

(4.7)

The definition of a double groupoid must be completed by the verification of a compatibility condition between the two structures, namely,

$$(\mathbb{A} \odot_H \mathbb{B}) \odot_V (\mathbb{C} \odot_H \mathbb{D}) = (\mathbb{A} \odot_V \mathbb{C}) \odot_H (\mathbb{B} \odot_V \mathbb{D}), \tag{4.8}$$

whenever the operations are possible. Graphically, this condition means that, in a square made of four squares whose edges in contact are matched, the result is the same whether one first composes horizontally and then vertically, or vice versa. In other words, the large square shown below makes sense.

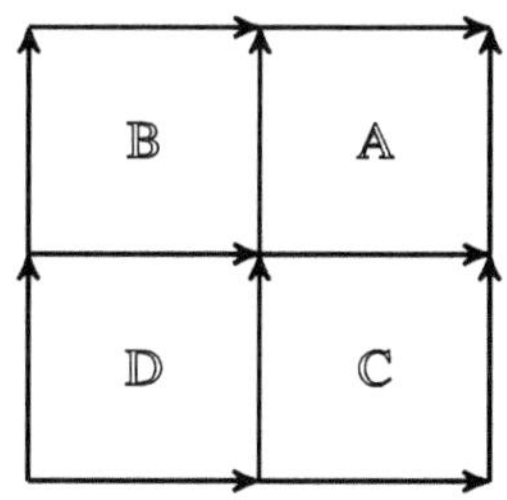

(4.9)

Exercise 4.4 Carefully verify that the compatibility condition embodied in Eq. (4.8) is satisfied. Make sure that in the diagram (4.9) the common edges guarantee that the operations are possible.

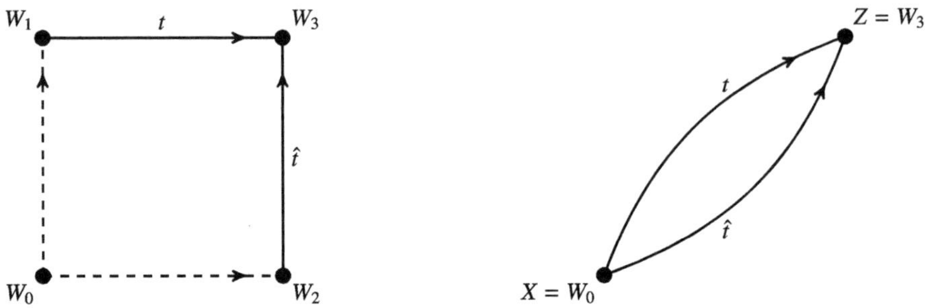

Fig. 4.3 An element in the core, where identity *arrows* are *dashed*

Finally, the *double source map* assigning to each square $\mathbb{A}$ the pair of its sources in the horizontal and vertical structures, $\tilde{\alpha}_H(\mathbb{A}), \tilde{\alpha}_V(\mathbb{A})$, is often assumed to be surjective. The graphical representation of this assumption is that a right angle made of a horizontal and a vertical arrow issuing from a common corner in $\mathcal{B}$, can always be completed (not necessarily uniquely) to a square in $\mathcal{Q}$. This is known as the *filling condition*.

A useful example of a double groupoid is the *coarse double groupoid*, $\square(\mathcal{H}, \mathcal{V})$, generated by two groupoids, $\mathcal{H}$ and $\mathcal{V}$, with a common base set. It consists of all the possible consistent squares that can be formed using the two groupoids as sides. Any double groupoid can be naturally mapped as an inclusion into the coarse double groupoid generated by its side groupoids.

Definition 4.1 The *core* of a double groupoid consists of the collection of all squares $\mathbb{A}$ for which the two source arrows s and $\hat{s}$ are, respectively, the units of $\mathcal{H}$ and $\mathcal{V}$ at the point $X = \alpha_H\left(\tilde{\alpha}_V(\mathbb{A})\right) = \alpha_V\left(\tilde{\alpha}_H(\mathbb{A})\right)$.

A square in the core groupoid, therefore, will have three coincident vertices, $W_1 = W_2 = W_3$, as represented in Fig. 4.3. For this reason, the core can be given a canonical groupoid structure by regarding its arrows as pairs of arrows $(t, \hat{t})$ with common source $X = W_0$ and target point $Z = \beta_H\left(\tilde{\beta}_V(\mathbb{A})\right) = \beta_V\left(\tilde{\beta}_H(\mathbb{A})\right) = W_3$. In many respects the core groupoid characterizes the double groupoid [6], although the physical identity of the original side groupoids is not preserved when passing to the core.

Example 4.1 (The Pair Double Groupoid) The Cartesian product $\mathcal{B}^4 = \mathcal{B} \times \mathcal{B} \times \mathcal{B} \times \mathcal{B}$ has a natural double groupoid structure, which can be identified with the coarse groupoid $\square(\mathcal{B}^2, \mathcal{B}^2)$. The two side groupoids are equal to the pair groupoid introduced in Example 1.3.

Exercise 4.5 Show a typical square $\mathbb{A}$ of the pair double groupoid and describe the effect of the source and target projections of the horizontal and vertical groupoid structures.

The literature on double groupoids is confined to high-level mathematical articles (e.g., [2, 6, 9, 23]). Physical applications do not abound, with the possible exception of quantum mechanics. In the more mundane context of classical continuum mechanics, however, it is not difficult to discern the natural appearance of double groupoids in such applications as solid binary mixtures. Chapter 5, therefore, is devoted entirely to this application.

4.2.1 Recapitulation and Foresight

The concepts introduced in this chapter may be summarized in the following more compact definition of a double groupoid.

Definition 4.2 A *double groupoid* (or *twofold groupoid*) is a set Q equipped with two groupoid structures whose bases are themselves ordinary groupoids (onefold groupoids), $\mathcal{Z}_1$ and $\mathcal{Z}_2$, with a common base $\mathcal{B}$. Denoting the compositions of the first and second groupoid structures of Q by $\odot_1$ and $\odot_2$, respectively, the following *compatibility condition* must be satisfied:

$$(\mathbb{A}\odot_1\mathbb{B})\odot_2(\mathbb{C}\odot_1\mathbb{D}) = (\mathbb{A}\odot_2\mathbb{C})\odot_1(\mathbb{B}\odot_2\mathbb{D}), \qquad \forall\, \mathbb{A}, \mathbb{B}, \mathbb{C}, \mathbb{D} \in Q, \tag{4.10}$$

whenever the operations are possible.

This compact definition turns out to incorporate all the elements meticulously, and somewhat haphazardly, introduced in the previous sections of the present chapter. To see that this is the case, starting from this definition alone, let us consider the meaning of a groupoid structure in Q whose base is an ordinary groupoid such as $\mathcal{Z}_1$. Since the elements of $\mathcal{Z}_1$ are arrows, the elements of Q connect two arrows (a source arrow and a target arrow) in $\mathcal{Z}_1$. A similar reasoning applies, mutatis mutandis, to the second groupoid structure of Q. What this means is that a typical element of Q involves four arrows, two from each of the groupoids $\mathcal{Z}_1$ and $\mathcal{Z}_2$. Accordingly, we designate elements of Q as *squares*. Moreover, the laws of composition of these squares must satisfy the compatibility condition, which strongly suggests to define these laws by considering the connections between a pair of arrows of $\mathcal{Z}_1$ as a pair of arrows of $\mathcal{Z}_2$, and vice versa. In this way, the compositions in Q can be defined in terms of the compositions in $\mathcal{Z}_1$ and $\mathcal{Z}_2$.

Looking ahead, in order to consider the possibility of generalizing these ideas inductively, we have also added a new element to our original description, an element that we have called the *standard square cell* $\mathcal{S}_2$ with vertices constituting a 4-tuple $\mathcal{V} = \{V_0, V_1, V_2, V_3\}$. A (not necessarily injective) map of $\mathcal{V}$ into $\mathcal{B}^4$ constitutes nothing but a selection of four (not necessarily distinct) points in the base manifold $\mathcal{B}$, that is, a 4-tuple $\mathcal{W} = \{W_0, W_1, W_2, W_3\}$. We can now regard a square in Q as a selection of four arrows (if any): two from $\mathcal{Z}_1$ (one with source W_0 and target W_2, and another with source W_1 and target W_3) and two from $\mathcal{Z}_2$ (one with source W_0 and target W_1, and another with source W_2 and target W_3.

In an n-fold groupoid, to be further defined in later chapters, the standard square cell (which can also be called a 2-cube) will be replaced with a standard n-cube cell (that is, a hypercube in n-dimensions). These cells have obvious inductive properties. For example, the standard 3-cube cell, is an ordinary cube with unit sides. It can be conceived in three different ways as a pair of parallel squares separated (in a three-dimensional space) by a unit distance and joined at the vertices by perpendicular edges. Similarly, the standard n-cube can be conceived in n different ways as a pair of parallel $(n-1)$-cubes separated (in an n-dimensional space) by a unit distance and joined at the vertices by perpendicular edges.

These inductive properties of hypercubes can be surmised as being portable to the (at this point conjectural or premonitory) definition of an n-fold groupoid on the basis of n given $(n-1)$-fold groupoids.

Definition 4.3 An n-fold groupoid is a set Q equipped with n groupoid structures whose bases are n given $(n-1)$-fold groupoids. Denoting the compositions of these groupoid structures of Q by $\odot_I$ $(I = 1, \ldots n)$, the following *compatibility condition* must be satisfied for every pair of indices $I \neq J$:

$$(\mathbb{A}\odot_I\mathbb{B})\odot_J(\mathbb{C}\odot_I\mathbb{D}) = (\mathbb{A}\odot_J\mathbb{C})\odot_I(\mathbb{B}\odot_J\mathbb{D}), \qquad \forall\, \mathbb{A}, \mathbb{B}, \mathbb{C}, \mathbb{D} \in Q, \tag{4.11}$$

whenever the operations are possible.

Remark 4.1 As we shall see in Chap. 7, Definition 4.3 needs to be finessed by stipulating that each of the structures involved can be ultimately traced back to a combination of n ordinary groupoids with a common base.

Exercise 4.6 Prove that Eq. (4.11) implies higher-order commutations. For example,

$$\begin{aligned} &((\mathbb{A}\odot_I\mathbb{B})\odot_J(\mathbb{C}\odot_I\mathbb{D}))\odot_K((\mathbb{M}\odot_I\mathbb{N})\odot_J(\mathbb{P}\odot_I\mathbb{Q}))\\ &=((\mathbb{A}\odot_K\mathbb{M})\odot_J(\mathbb{C}\odot_K\mathbb{P}))\odot_I((\mathbb{B}\odot_K\mathbb{N})\odot_J(\mathbb{D}\odot_K\mathbb{Q}))\,. \end{aligned} \tag{4.12}$$

Chapter 5
Binary Composites

Abstract A composite is a material in which two distinct constituents are discernible at the microscopic or mesoscopic scale. A familiar example is represented by layered materials such as plywood, but much more sophisticated examples appear in nature or are produced by modern techniques of additive manufacturing. Since each of the constituents is a material in its own right, it should not be surprising that for a binary composite a double groupoid arises naturally as a faithful mathematical representation of its local and distant symmetries. Even if the two constituents happen to be individually uniform, the composite may not be so. In this case, one says that the composite is misaligned or that it contains a distribution of defects. This circumstance can be quantified in terms of the properties of the double groupoid.

5.1 Motivation

Most materials, both natural and artificial, when examined at a microscopic or mesoscopic scale, reveal the coexistence of different constituent structures, whose combination results in the observable macroscopic behaviour. According to context or scope of application one can speak about composites, mixtures, additive manufacturing, or metamaterials. Each constituent of any of these complex media can be considered as a material in its own right. Nevertheless the technique of *homogenization*, widely used in the analysis of composites, provides sophisticated approximations to a putative single material equivalent in its mechanical response to the original composite.

To illustrate what may be lost with a technique of homogenization of a composite, consider a two-dimensional universe in which two sheets of material have been glued together throughout their surface, just as it is done to create plywood from two (theoretically infinitesimally thin) sheets of wood, as shown in Fig. 5.1. With techniques of 3D printing and metamaterials, the same situation can be conceived in the real three-dimensional world, but we will stay in two dimensions for the sake of clarity of the exposition.

The two plies are made of the same material, but one of the plies, shown on the right side of Fig. 5.1, has been deformed (by some forces, couples, or thermal

© The Author(s), under exclusive license to Springer Nature Singapore Pte Ltd. 2026
M. Epstein, *An Elementary Introduction to* n*-fold Groupoids*,
Surveys and Tutorials in the Applied Mathematical Sciences 19,
https://doi.org/10.1007/978-981-95-1869-2_5

Fig. 5.1 Two thin plies of wood before gluing

gradients applied in its plane) before the gluing operation takes place. The glued assembly is then released whereupon the misalignment between the two structures results in the appearance of residual stresses in each layer. If instead of two inert structures we had combined two biological membranes in a similar way, the appearance of a difference of chemical potential between the layers could have resulted in a process of *time evolution* whereby some state of equilibrium is negotiated in terms of structural changes in both layers to compensate for the initial misalignment. It is precisely this possible misalignment that would be interesting to quantify.

5.1.1 Measures of Misalignment

Recall (Fig. 3.5) our schematic representation of a uniform body made of a single constituent. Along the same lines, a schematic representation of a binary composite whose constituents happen to be uniform, is shown in Fig. 5.2. Each of the uniform constituents is characterized in its material response by a material archetype and its implants at each point of the composite in a reference configuration. We ask ourselves: is this composite body materially uniform? Does it matter, and if so,

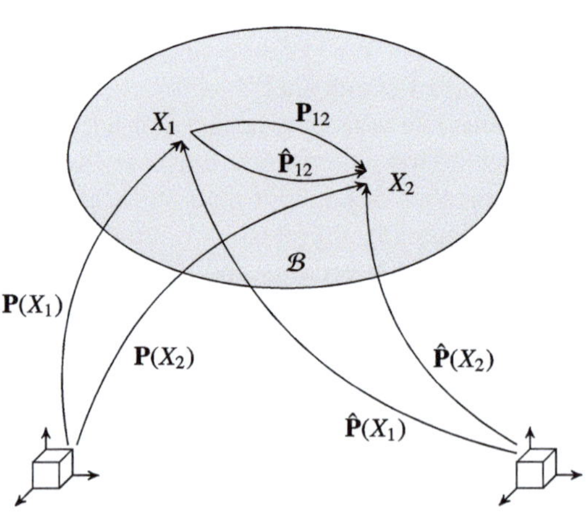

Fig. 5.2 A binary composite and its two archetypes

how? We will show that if the composite fails to be uniform, we can recognize a new kind of defectivity not present in ordinary simple materials.

As briefly described in Sect. 3.2.2, the presence of defects in a simple (single constituent) material assumes that the body is already uniform, while the defects manifest themselves as the inability of the body to find a configuration whereby the transplants are mere translations. In the case of a binary composite, however, even when both constituents are individually uniform (or even homogeneous), if the mixture happens not to be uniform we have a situation of a defect of *misalignment* between the constituents. The quantification of the density of these defects, not present in simple materials, turns out to depend on the interplay between the archetypal symmetry groups of the constituents.

Two Triclinic Constituents

As described in Chap. 3, a triclinic material has no non-trivial symmetries, that is, no material symmetries except the identity. This being the case, the implant maps $\mathbf{P}(X)$ and $\hat{\mathbf{P}}(X)$ of both constituents are uniquely defined once the two archetypes have been chosen. We conclude that the composite is materially uniform if, and only if, the transplant maps $\mathbf{P}_{12}$ and $\hat{\mathbf{P}}_{12}$, shown in Fig. 5.2, coincide for every pair of points X_1 and X_2.

Since

$$\mathbf{P}_{12} = \mathbf{P}(X_2)\ (\mathbf{P}(X_1))^{-1} \tag{5.1}$$

and

$$\hat{\mathbf{P}}_{12} = \hat{\mathbf{P}}(X_2)\ \left(\hat{\mathbf{P}}(X_1)\right)^{-1}, \tag{5.2}$$

we may state the following:

Theorem 5.1 *A binary composite made of uniform elastic constituents with no non-trivial symmetries is uniform if, and only if, the matrix field*

$$\mathbf{A}(X) = \left(\hat{\mathbf{P}}(X)\right)^{-1}\ \mathbf{P}(X) \tag{5.3}$$

is constant.

A Triclinic and an Isotropic Constituent

Let the archetype of the isotropic constituent be in a natural stress-free configuration. The archetypal symmetry group of this (hatted) constituent is, therefore,

$$\hat{\mathcal{G}} = \mathcal{O}_3. \tag{5.4}$$

Let $\hat{\mathbf{P}}(X)$ be an implant of this archetype into the point $X \in \mathcal{B}$, and let $\mathbf{Q}$ be an arbitrary proper orthogonal 3×3 matrix. Then, the product $\hat{\mathbf{P}}(X)\,\mathbf{Q}$ is also a valid implant at the same point. It follows from here that the most general transplant between points X_1 and X_2 of the uniform isotropic constituents is of the form

$$\hat{\mathbf{P}}_{12} = \left(\hat{\mathbf{P}}(X_2)\,\mathbf{Q}_2\right)\,\left(\hat{\mathbf{P}}(X_1)\,\mathbf{Q}_1\right)^{-1} \tag{5.5}$$

where $\mathbf{Q}_1$ and $\mathbf{Q}_2$ are arbitrary proper orthogonal matrices. For the triclinic constituent the transplant is uniquely defined by Eq. (5.1). Uniformity of the composite requires, therefore, for every pair of points, the existence of proper orthogonal matrices $\mathbf{Q}_1$ and $\mathbf{Q}_2$ such that

$$\left(\hat{\mathbf{P}}(X_2)\,\mathbf{Q}_2\right)\,\left(\hat{\mathbf{P}}(X_1)\,\mathbf{Q}_1\right)^{-1} = \mathbf{P}(X_2)\,\left(\mathbf{P}(X_1)\right)^{-1}. \tag{5.6}$$

Using the definition of the matrix field $\mathbf{A}(X)$ in Eq. (5.3), condition (5.6) can be rewritten as

$$\mathbf{R}\,\mathbf{A}(X_1) = \mathbf{A}(X_2), \tag{5.7}$$

where $\mathbf{R} = \mathbf{Q}_2\,\mathbf{Q}_1^T$ is a proper orthogonal matrix. Multiplying each side of this equation by its transpose, we conclude that uniformity implies that the field of symmetric matrices defined as

$$\mathbf{B}(X) = \mathbf{A}^T(X)\,\mathbf{A}(X) \tag{5.8}$$

must be a constant field over the body.

Conversely, let the field $\mathbf{B}(X)$, defined by two implant fields $\mathbf{P}(X)$ and $\hat{\mathbf{P}}(X)$, be constant. Then, for any two given body points, X_1 and X_2, we can write

$$\mathbf{A}^T(X_2)\,\mathbf{A}(X_2) = \mathbf{A}^T(X_1)\,\mathbf{A}(X_1). \tag{5.9}$$

This equation can be rewritten as

$$\left((\mathbf{A}(X_1))^{-1}\right)^T\,\mathbf{A}^T(X_2) = \mathbf{A}(X_1)\,\left(\mathbf{A}(X_2)\right)^{-1}, \tag{5.10}$$

or

$$\left(\mathbf{A}(X_2)\,(\mathbf{A}(X_1))^{-1}\right)^{-1} = \left(\mathbf{A}(X_2)\,(\mathbf{A}(X_1))^{-1}\right)^T. \tag{5.11}$$

Hence the matrix $\mathbf{A}(X_2)\,(\mathbf{A}(X_1))^{-1}$ is orthogonal and we recover condition (5.7). This completes the proof of the following:

Theorem 5.2 *A binary composite made of uniform elastic constituents, one of which is triclinic and the other one isotropic, is uniform if, and only if, the symmetric matrix field*

$$\mathbf{B}(X) = \mathbf{A}^T(X)\,\mathbf{A}(X) \tag{5.12}$$

with

$$\mathbf{A}(X) = \left(\hat{\mathbf{P}}(X)\right)^{-1}\,\mathbf{P}(X) \tag{5.13}$$

is constant.

Remark 5.1 If the constituents are *smoothly uniform* (so that their implant fields can be chosen to be smooth over $\mathcal{B}$), we can define a *local* measure of misalignment as the gradient of the tensor field $\mathbf{A}$ or $\mathbf{B}$, respectively, for the two cases discussed above.

Remark 5.2 For more general cases of two uniform constituents with various symmetry groups, similar criteria can be obtained for the composite to be uniform. These criteria have been studied at length elsewhere [11].

Exercise 5.1 Find a necessary and sufficient condition for a composite made of two uniform and isotropic constituents to be uniform.

5.2 Construction and Meaning of the Material Double Groupoid

In the general case of materially uniform constituents, the groupoids $\mathcal{Z}_1 \rightrightarrows \mathcal{B}$ and $\mathcal{Z}_2 \rightrightarrows \mathcal{B}$ of the constituents are transitive subgroupoids of the 1-jet groupoid $\Pi^1(\mathcal{B}, \mathcal{B})$ introduced in Sect. 3.1.6. We can, accordingly, form the coarse double groupoid $\square(\mathcal{Z}_1, \mathcal{Z}_2)$ defined in Sect. 4.2. Out of this large generic double groupoid, whose squares include all possible pairs of arrows, one pair from each groupoid, joining every ordered set of four points of $\mathcal{B}$, we would like to select only those squares that convey a physical property of a binary composite. The following definition achieves this objective.

Definition 5.1 The material double groupoid of a binary composite

$$\begin{array}{ccc} \mathcal{Q} & \rightrightarrows & \mathcal{Z}_2 \\ \downdownarrows & & \downdownarrows \\ \mathcal{Z}_1 & \rightrightarrows & \mathcal{B} \end{array} \tag{5.14}$$

consists of all *commutative squares* of $\square(\mathcal{Z}_1, \mathcal{Z}_2)$.

Remark 5.3 A square

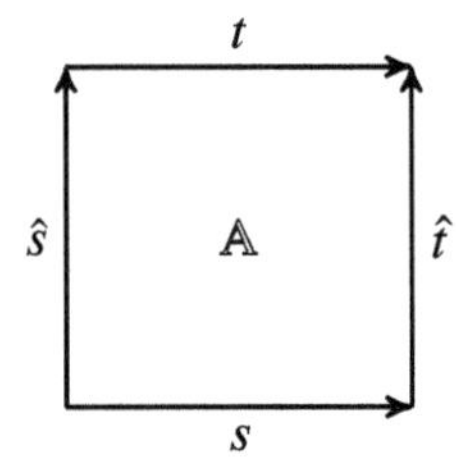

(5.15)

is said to be *commutative* if

$$\hat{t}\, s = t\, \hat{s}. \tag{5.16}$$

It is important to emphasize that this equality is only possible when, as is the case in a material double groupoid, both side groupoids are subgroupoids of a common groupoid. Otherwise, there is no meaning that can be attributed to the composition of arrows in Eq. (5.16). In our case, the common underlying groupoid is the 1-jet groupoid $\Pi^1(\mathcal{B}, \mathcal{B}) \rightrightarrows \mathcal{B}$.

Exercise 5.2 Verify that in the pair double groupoid all squares are automatically commutative.

To unravel the physical interpretation of the commutativity of a square, let us consider the horizontal arrows s and t. Their sources are indicated in the diagram (5.17) below as X_0, X_1, and their targets as X_2, X_3.

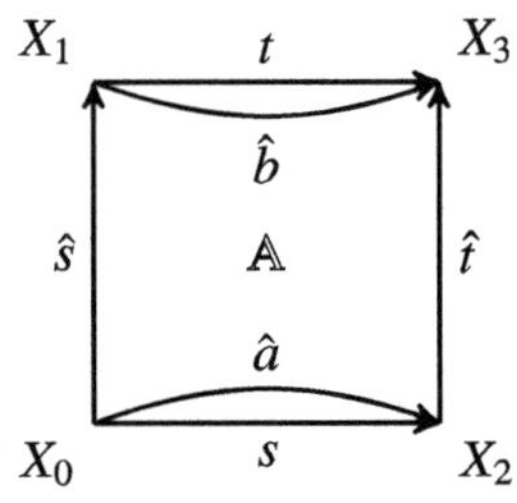

(5.17)

Since the constituents have been assumed to be uniform, we may choose an arrow $\hat{a}$ belonging to the vertical side groupoid $\mathcal{Z}_2$, as shown in the diagram (5.17). In the same diagram, moreover, we have added the arrow $\hat{b} = \hat{t}\hat{a}\hat{s}^{-1}$, also belonging to $\mathcal{Z}_2$. The arrows $\hat{a}$ and $\hat{b}$ are not part, in general, of a square of the double groupoid, but they obviously exist (not necessarily uniquely) since $\mathcal{Z}_2$ is transitive. We define a loop-shaped arrow in $\Pi^1(\mathcal{B}, \mathcal{B})$ at X_0 as

$$m_0 = \hat{a}^{-1}\, s. \tag{5.18}$$

We refer to m_0 as the *horizontal alignment* of $\hat{a}$ with s. Similarly, at X_1 we define the corresponding alignment of $\hat{b}$ with t as

$$m_1 = \hat{b}^{-1}\, t. \tag{5.19}$$

We remark that a unity alignment, namely $m_0 = \epsilon_{\mathcal{Z}_1}(X_0)$, corresponds to the existence of a common arrow between X_0 and X_2 for the two material groupoids. In other words, X_0 and X_2, considered as points of the composite, would be materially isomorphic.

The following theorem is a direct consequence of the definitions just stated.

Theorem 5.3 *A square is commutative (in accordance with Eq. (5.16)) if, and only if, the alignment* m_1 *is the conjugate of* m_0 *via* $\hat{s}$*, that is,*

$$m_1 = \hat{s}\, m_0\, \hat{s}^{-1}. \tag{5.20}$$

A similar theorem holds true for the *vertical alignments* $\hat{m}_0 = a^{-1}\,\hat{s}$ and $\hat{m}_2 = b^{-1}\,\hat{t}$, where a is an arrow in the horizontal side groupoid and b is given by $b = t\,a\,s^{-1}$, as suggested in the diagram (5.21).

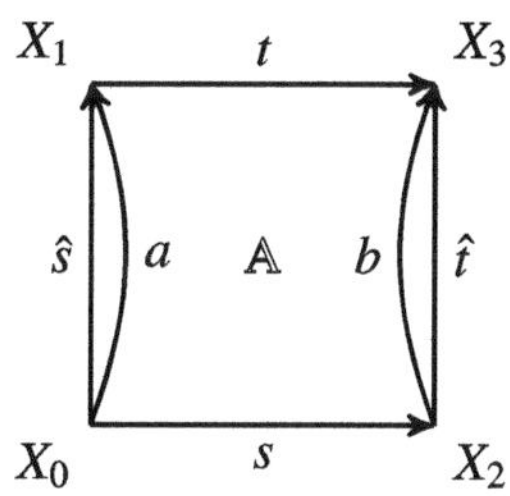

(5.21)

The conjugacy condition is now

$$\hat{m}_2 = s\,\hat{m}_0\, s^{-1}. \tag{5.22}$$

Going back to our conceptual definition of a double groupoid in Sect. 4.1, what can come closer to exemplifying the role of a pair of arrows of one groupoid as hyper-arrows between arrows of another? When a square is commutative the vertical arrows indicate that corresponding horizontal alignments are conjugate, while the horizontal arrows indicate that corresponding vertical alignments are conjugate!

Remark 5.4 The relation of conjugacy within a group is usually an indication of equivalence of meaning. In the case of the general linear group, whose elements are non-singular square matrices, conjugacy is known as *similarity*. Two similar matrices share the same characteristic polynomial. Thus, they can be regarded as *the same linear operator* expressed in two different bases. In solid mechanics, moreover, conjugacy of two quantities arises naturally upon changes of reference configuration of the body.

Fig. 5.3 Artistic rendition of the building block of a double groupoid

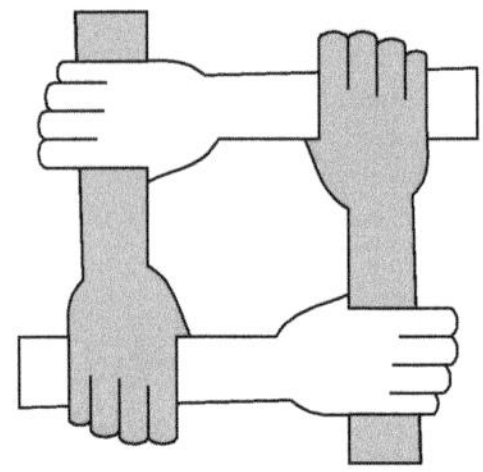

We may say that the property of commutativity of a square introduces a new kind of symmetry. Indeed, local symmetries at a point are represented by groups. Distant symmetries between two points manifest themselves as arrows of a groupoid. The squares of a double groupoid can be regarded, therefore, as shared symmetries between two pairs of points. The physical meaning of these shared symmetries in the case of a binary composite consists of having 'the same' degree of alignment between the arrows of both underlying materials. An artistic rendition of this shared property is attempted in Fig. 5.3.

5.2.1 *The Core Condition*

According to Definition 4.1 the core of a double groupoid consists of squares of the form

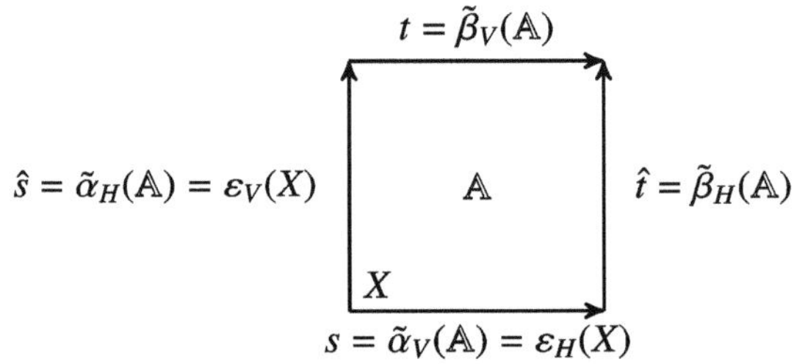

(5.23)

By the very meaning of the identities, this kind of square can also be represented as

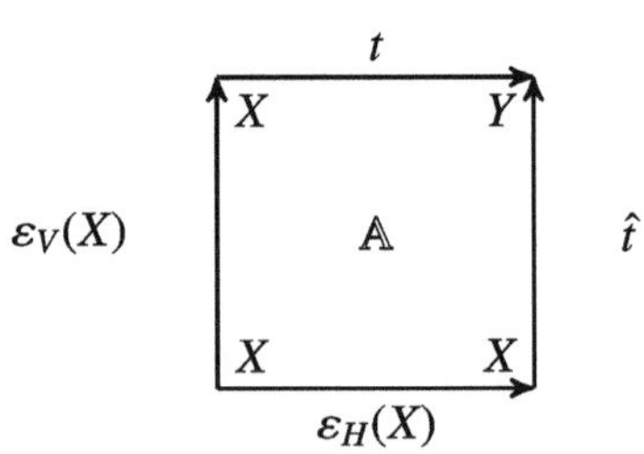

(5.24)

Since our double groupoid is made of commutative squares, we obtain the following result.

Theorem 5.4 *A composite made of two materially uniform constituents is uniform if, and only if, the core groupoid is transitive.*

Indeed, the commutativity of a square in the core implies that $\hat{t} = t$. Recall that the core groupoid is an ordinary groupoid each of whose arrows consists of a pair $(t, \hat{t})$. The transitivity of this groupoid means that for every $X, Y \in \mathcal{B}$ there is a common material isomorphism for both constituent materials.

Corollary 5.1 *The core groupoid of the material double groupoid of a composite made of a triclinic constituent and an isotropic constituent, both of which are materially uniform, is transitive if, and only if, the symmetric tensor field* $\mathbf{B}(X)$ *defined in Eq. (5.8) is constant over the body.*

This corollary is a direct consequence of Theorems 5.2 and 5.4.

Remark 5.5 (Measures of Lack of Transitivity) Corollary 5.1 (and similar ones for other symmetry classes) can be given an interesting interpretation. Assuming that fields such as $\mathbf{A}(X)$ or $\mathbf{B}(X)$ are differentiable, their gradients are legitimate measures of the local lack of uniformity of a composite. In terms of the underlying double groupoid, this somewhat unexpected result (possibly meaningful in other applications) provides a *local measure of intransitivity* of the core of certain double groupoids having two transitive side groupoids.

5.2.2 Designing Distributions of Defects of Misalignment in Metamaterials

Recent research (e.g. [1, 25]) reveals that intentionally designed distributions of topological defects in metamaterials can have a beneficial effect in steering stresses and deformations. In terms of the continuous limit, this kind of design can be represented by, for example, a doubly periodic planar field $\mathbf{A}(X)$ or $\mathbf{B}(X)$ in some reference configuration. This deliberately introduced defectivity of misalignment between several constituents of a metamaterial will have repercussions in the solution of the associated boundary value problem in elasticity.

Example 5.1 (Periodic Distribution of Defects of Misalignment) A planar body consists of two uniform constituents made of the same orthotropic material, the only difference being the respective implant fields. As in Example 3.2, a stress-free archetype of a solid can be regarded as the orthonormal natural basis $\mathbf{E}_\alpha$ $(\alpha = 1, 2)$ of $\mathbb{R}^2$. A field of implants $\mathbf{P}(X)$ for a uniform body is represented as a field of bases $\mathbf{P}_\alpha$ $(\alpha = 1, 2)$ given in Cartesian components by

$$\mathbf{P}_\alpha = P^I_\alpha \, \mathbf{E}_I, \tag{5.25}$$

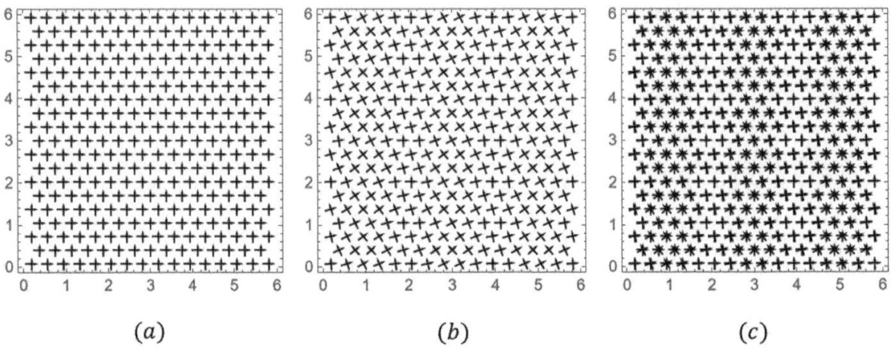

Fig. 5.4 Lack of uniformity due to misalignment of uniform constituents: (a) Constituent 1; (b) Constituent 2; (c) Composite 1+2

where we are adhering to the summation convention for diagonally repeated indices. For the first constituent, it will be assumed that the field of matrices $[P](X)$ is everywhere equal to the unit matrix $[I]$. The corresponding field of bases is represented in Fig. 5.4a. For the second constituent, made of the same orthotropic material, the implant field has been deliberately modulated according to the rotation matrix

$$[\hat{P}] = \begin{bmatrix} \cos\theta(x, y) & -\sin\theta(x, y) \\ \sin\theta(x, y) & \cos\theta(x, y) \end{bmatrix}, \tag{5.26}$$

with

$$\theta(x, y) = 0.5 \sin^2(\pi\, x/2) + 0.3 \sin^2(\pi\, y). \tag{5.27}$$

This field of bases is represented in Fig. 5.4b. The superposition of both fields is shown in Fig. 5.4c. Although both constituents are individually uniform, the resulting composite is not. The material has been tuned to perform in a desired way suited to the application at hand, or the misalignment between the constituents may be due to a fabrication defect. We remark the essential difference between defects of inhomogeneity in a simple uniform material and defects of misalignment in a composite. Even if the two constituents were not just uniform but also homogeneous, a misalignment could happen as the result of initial thermal strains (or other causes) in one of the constituents at the time of assembly.

Exercise 5.3 Obtain the explicit expression for the (tensorial) local measure of misalignment in the previous example. Show that this measure is periodic in both x and y and that it vanishes at the points $(x, y) = (j\pi, k\pi/2)$ for all integers j and k. [Hint: obtain the gradient of the tensor $\mathbf{A}$.]

Chapter 6
Hypercubes and Skeletons

Abstract The elements of a double groupoid Q can be regarded as squares made up of two pairs of arrows, each pair being elicited from one of the base groupoids. With a view to generalizing this fertile idea, it is convenient to review the properties of hypercubes, that is, cubes in an arbitrary number of dimensions. A hypercube in dimension n contains faces of all dimensions smaller than n. The 0-faces are called vertices, and the 1-faces are called edges. The skeleton of a hypercube is the collection of its edges and vertices, the 'flesh' having been discarded. Skeletons can be regarded as directed graphs by adopting an ingenious binary notation for its vertices. By the same device, the vertices of a hypercube can be organized into a useful and natural hierarchical structure.

6.1 Why Hypercubes?

We have by now gathered enough experience to be able to surmise that a generalization of the concept of groupoid can perhaps be rooted in the relation between certain paradigmatic geometric entities (a unit segment, a unit-side square) and the properties shared by some objects somehow attached to their vertices. This relation suggests that one of the possible avenues to introduce and understand the concept of n-fold groupoid might arise, as a point of departure, from a generalization of the said one- and two-dimensional geometric entities to higher dimensional counterparts.

Definition 6.1 The solid n-dimensional unit hypercube Q_n (or n-cube for short) is the geometric object defined as the Cartesian product

$$Q_n = \underbrace{J \times \ldots \times J}_{\text{n times}} = J^n \subset \mathbb{R}^n, \tag{6.1}$$

where $J = [0, 1]$ is the oriented closed unit segment in $\mathbb{R}$.

Remark 6.1 According to this definition, Q_1 is the oriented unit segment, and Q_2 is indeed the square with oriented unit sides represented in Fig. 4.1. The realization

© The Author(s), under exclusive license to Springer Nature Singapore Pte Ltd. 2026
M. Epstein, *An Elementary Introduction to* n*-fold Groupoids*,
Surveys and Tutorials in the Applied Mathematical Sciences 19,
https://doi.org/10.1007/978-981-95-1869-2_6

of these entities resulted in the basic building blocks of groupoids and double groupoids, respectively.

For each integer $0 \leq h < n$, an *h-face* F_h is a subset of Q_n obtained by fixing $n - h$ coordinates of $\mathbb{R}^n$ to either 0 or 1. For example, a 4-face of Q_7 might be the set

$$F_4 = \{0\} \times J \times \{1\} \times \{0\} \times J \times J \times J. \tag{6.2}$$

Every h-face of Q_n is congruent (by a Euclidean isometry) with the standard unit h-cube Q_h. For a fixed h, the number of h-faces is obtained as

$$N_h = 2^{n-h} \binom{n}{h} = 2^{n-h} \frac{n!}{h!\,(n-h)!}. \tag{6.3}$$

Exercise 6.1 Prove Eq. (6.3). [Hint: In how many ways can $n - h$ elements be chosen out of a set of n elements? Further, in how many different ways can the digits 0 and 1 be arranged in a sequence of length $(n - h)$?]

A 0-face of Q_n is called a *vertex*. The number of vertices is, therefore, $N_0 = 2^n$. A 1-face is called an *edge*. The number of edges of Q_n is $N_1 = 2^{n-1}n$. Thus, for example, the three-dimensional cube has 12 edges. The edges of Q_n can be partitioned into n classes, according to the coordinate axis of $\mathbb{R}^n$ to which they are parallel (namely, the running coordinate, since all other coordinates have been fixed to a value of either 0 or 1). If we denote the natural coordinates of $\mathbb{R}^n$ by X_I (with $I = 1, \ldots, n$) we can loosely talk about the class of X_I-edges. For the ordinary unit cube in three dimensions (with coordinates X, Y, Z), we have four X-edges, four Y-edges, and four Z-edges.

An important relation between vertices is that of *adjacency*. Recalling that all vertices of Q_n are points with coordinates in $\mathbb{R}^n$, each of which has a value 0 or 1, two vertices of Q_n are said to be *adjacent with each other* if, and only if, their respective coordinates are identical except for a single entry. It follows from this definition that each vertex is adjacent with exactly n other vertices. Two edges are said to be adjacent if they have a common vertex. We remark that adjacency is not an equivalence relation since it is neither reflexive nor transitive. Since, by definition, an edge is obtained by fixing $n - 1$ coordinates, edges join only adjacent vertices. The class to which an edge belongs is precisely determined by the coordinate X_I in which the two adjacent vertices differ.

Exercise 6.2 Make sure that the assertions just made are true. In case of doubt, reaffirm your intuition by using the ordinary three-dimensional cube as an example.

At the other extreme, for $h = n - 1$, we obtain the so-called *facets* of the n-cube. The number of facets is $2n$. There are exactly two facets perpendicular to each of the coordinate axes. We may think of n classes of X_I-facets, each containing a pair of parallel facets. For example, the two X_2-facets of Q_7 are $J \times \{0\} \times J \times J \times J \times J \times J$ and $J \times \{1\} \times J \times J \times J \times J \times J$. Since each facet is congruent with Q_{n-1}, a

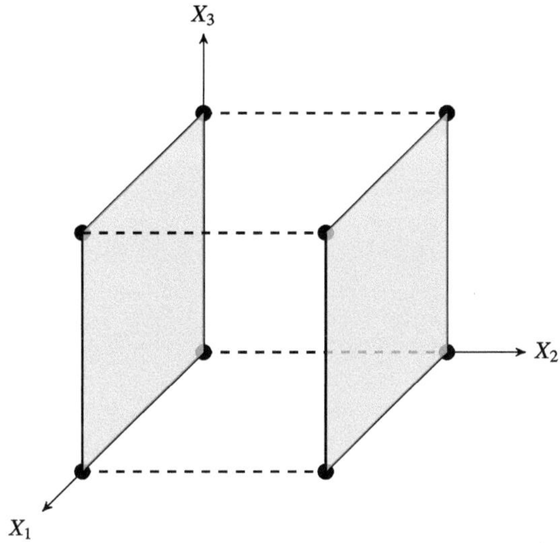

Fig. 6.1 Cube generated from two X_2-faces

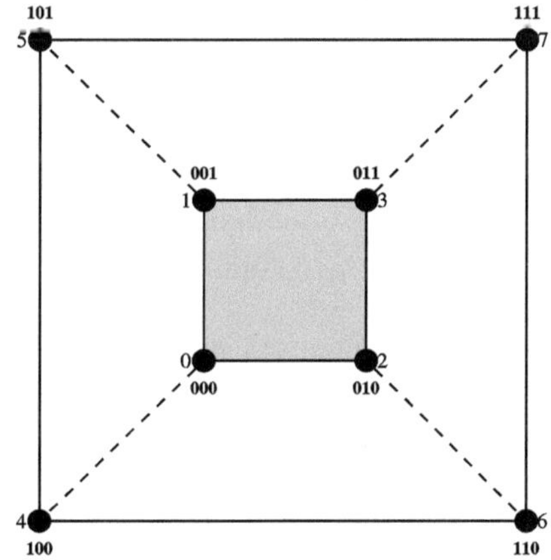

Fig. 6.2 Cube generated from two X_1-faces, in one-point perspective

creative and useful inductive way to think of Q_n is as two parallel $(n-1)$-cubes of class X_I in $\mathbb{R}^n$, initially coincident, one of which has swept a unit distance along X_I. The vertices of the moving facet trace the new edges of Q_n. In this way, we can think of the square as being comprised between two parallel segments joined at their corresponding ends, and of a cube as the volume comprised between two parallel squares joined at their corresponding vertices. Figure 6.1 shows how the ordinary cube is generated from two squares of class X_2, one of which has travelled a unit distance along X_2.

The same cube can be generated from two squares of class X_1 joined by four X_1-edges. For the sake of variety, Fig. 6.2 represents this generation process in one-point-perspective, whereby parallel lines appear to converge at a point, just as it appears to an observer when one of the two squares is receding along the first coordinate axis, like the back of a truck travelling along a road. The observer is

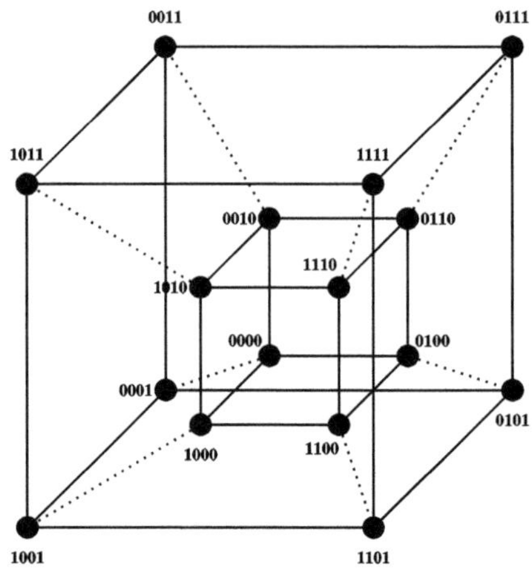

Fig. 6.3 Tesseract obtained by joining corresponding vertices of two cubes by means of edges running along the fourth dimension (*dotted lines*)

now placed far along the X_1-axis looking backwards toward the origin. The shaded square represents the face lying in the plane X_2, X_3.

In Fig. 6.2 the vertices of the cube have been labelled (above or below each vertex) by their respective coordinates X_1, X_2, X_3 in $\mathbb{R}^3$, without intermediate commas. This notation can be conveniently exploited to assign a single number to each vertex. Indeed, these sequences of zeros and ones can be interpreted as the binary representations of natural numbers, which can then be expressed in the decimal system. Adopting this policy, we have also shown this numbering by labels placed to the right or to the left of each vertex.

One can conceive the 4-cube (or *tesseract*) in a similar vein, imagining two ordinary cubes initially coincident, one of which is then subjected to a unit displacement into a putative fourth dimension, indicated with dotted lines in Fig. 6.3.[1] For clarity, the fourth dimension is rendered in one-point perspective, thus completing the tesseract with eight new edges (dashed in the figure) running parallel to the fourth coordinate axis and connecting the corresponding vertices of the cubes. The same binary-to-decimal numbering technique can be used for the tesseract, but the decimal labels have been omitted in the figure for clarity.

6.2 Skeletons and Graphs

The constructions just described can be made more precise by considering the *skeleton* S_n of the n-cube Q_n. The skeleton S_n is defined as the collection of vertices and edges of Q_n, occupying their positions in $\mathbb{R}^n$. The flesh, so to speak, has been removed, leaving just the bare bones. The skeleton S_n is also known as the *graph* of

[1] Adapted from https://tex.stackexchange.com/questions/598573/.

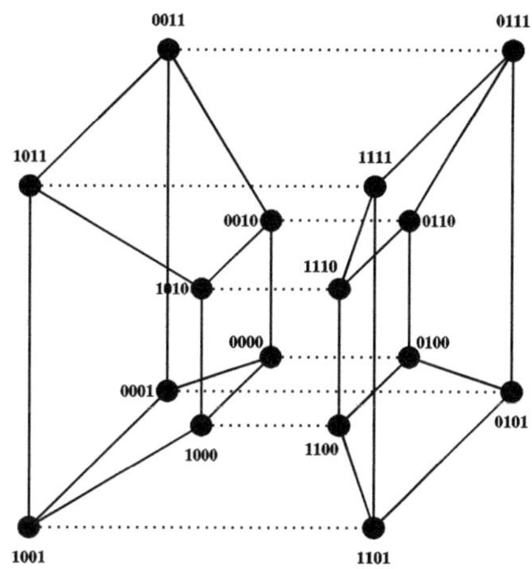

Fig. 6.4 The tesseract of Fig. 6.3 obtained by joining corresponding vertices of the two cubes shown (*solid lines*) by means of edges (*dotted lines*) running along the second coordinate axis

Q_n. A comprehensive knowledge of *graph theory* [17, 18] is not necessary for our application. In general terms, a *graph* is a finite set of points (or *vertices*) together with a set of lines (or *edges*), each joining two distinct vertices. Two vertices, A and B, joined by an edge e are said to be *adjacent* with each other, while e is said to be *incident* with A and with B. This terminology coincides so far with the terminology used above for n-cubes.

In terms of n-skeletons, rather than n-cubes, these inductive constructions consistently deliver skeletons of all orders. Indeed, the n-skeleton is obtained by joining two $(n-1)$-skeletons by means of perpendicular edges. Just as in the case of n-cubes, therefore, it is easy to see that the n-skeleton generated by a pair of parallel $(n-1)$-skeletons is independent of the class of the two facets of departure. Indeed, the skeleton of the tesseract of Fig. 6.3 can be also obtained, as suggested in Fig. 6.4, by joining two ordinary 3-skeletons (in the subspace with coordinates X_1, X_3, X_4) by means of edges running along X_2.

Definition 6.2 A *walk* in a graph is a sequence of adjacent vertices. A finite walk p with m steps, therefore, can be explicitly given as an $(m+1)$-tuple

$$p = \{v_0, v_1, \ldots, v_m\} \tag{6.4}$$

with the condition of adjacency between v_{i-1} and v_i for each $i = 1, \ldots, m$.

Definition 6.3 A walk is said to be *closed* if the first and last entries of the sequence p coincide.

Remark 6.2 In graph theory there is a need to distinguish between walks, trails, and paths, a distinction that will not be necessary at this point. In a walk, vertices and edges may be visited more than once along the way.

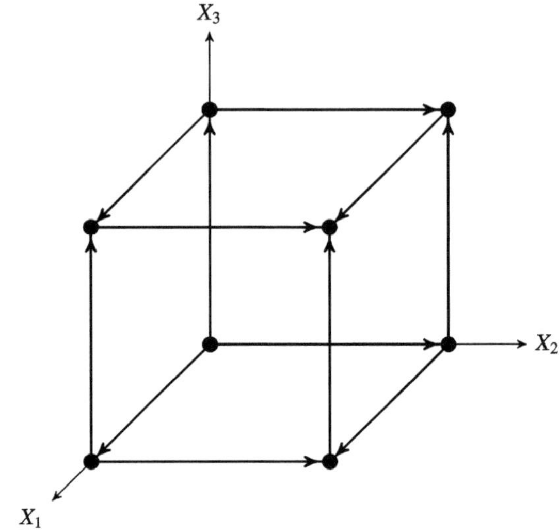

Fig. 6.5 The directed 3-skeleton S_3

6.2.1 Directed and Group-Weighted Skeletons

In our case, since we have considered an orientation of J, we obtain a *directed graph*, as shown in Fig. 6.5 for the case $n = 3$. In a directed graph every pair of adjacent vertices (that is, every edge) is considered as an ordered pair, the first element of the pair indicating the tail of the arrow.

For a directed graph the notion of adjacency can be refined. In the case of our directed skeletons, of two vertices adjacent with each other we declare that the vertex having an extra coordinate of value 1 is *adjacent from* the vertex having a corresponding 0 coordinate. Conversely, the latter vertex is said to be *adjacent to* the former. 'Adjacency from' is not even a symmetric relation. The oriented skeleton S_n consists, therefore, of the vertices and the oriented edges (unit vectors) joining each vertex to those which are adjacent from it. The number of oriented edges issuing from or arriving at each vertex is not constant. Thus, the vertex at the origin is not adjacent from any other vertex.

Definition 6.4 Let $\mathcal{G}$ be a group and let S be a graph whose collection of edges we denote by $\mathcal{E}$. We say that S is a *$\mathcal{G}$-weighted graph* if a function

$$w : \mathcal{E} \to \mathcal{G}, \tag{6.5}$$

called a *weight function* has been defined. This function need not be injective.

In a significant departure from conventional graph theory, in the case of directed graphs we will permit the consideration of all walks in the underlying non-directed graph. In the directed graph, however, the weight of an edge traversed against its

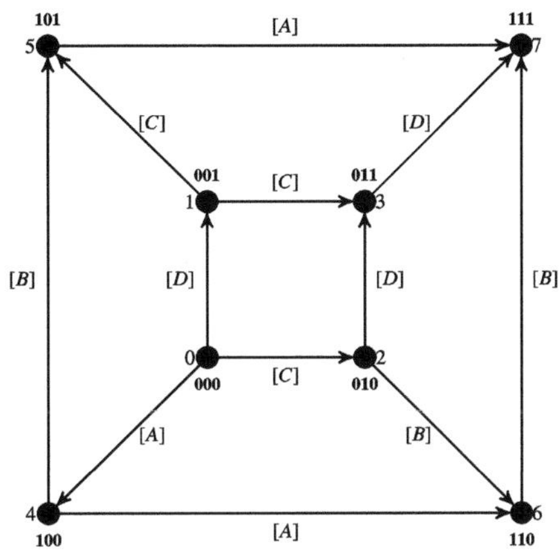

Fig. 6.6 A directed and group-weighted skeleton S_3

orientation will be replaced with the inverse of the weight assigned by the weight function w.

Definition 6.5 The *weight of a (finite) walk p* in a group-weighted and directed graph S is the group composition of the weights of the successive edges of the walk, provided inverse weights are assigned to edges traversed against the current.

For lack of a better term we will refer to this policy as the *direct-inverse* weighting convention. According to this definition, if a walk consists of a return to the vertex of departure via the reversal of the steps of a previous walk, the combination of both walks will have a total weight equal to the group identity.

Exercise 6.3 A directed and $GL(3, \mathbb{R})$-weighted 3-skeleton $\mathcal{S}_3$ is represented as the planar graph shown in Fig. 6.6. The weights are indicated in the figure as 3×3 matrices $[A]$, $[B]$, $[C]$, $[D]$ defined as

$$[A] = \begin{bmatrix} 2 & 0 & 0 \\ 0 & 1 & 0 \\ 0 & 0 & 1 \end{bmatrix}, \quad [B] = \begin{bmatrix} 0 & 0 & -1 \\ 0 & 1 & 0 \\ 1 & 0 & 0 \end{bmatrix}, \quad [C] = \begin{bmatrix} 0 & 0 & 1 \\ 0 & -1 & 0 \\ 1 & 0 & 0 \end{bmatrix}, \quad [D] = \begin{bmatrix} 0 & 0 & 1 \\ 0 & 1 & 0 \\ -1 & 0 & 0 \end{bmatrix}.$$

Evaluate the weight of the walks defined by the following sequences of vertices:

1. $\{0,4,5,1,0\}$,
2. $\{0,1,3,2,6,7,5,4,0\}$.

Exercise 6.4 What would the answers to Exercise 6.3 be if the additive group of matrices $\left(\mathbb{R}^{3\times 3}, +\right)$ were used instead of the multiplicative group $GL(3, \mathbb{R})$? Recall that under matrix addition the groups of matrices of the same order (same number of rows and same number of columns) are Abelian. What is the neutral element of $\left(\mathbb{R}^{m\times n}, +\right)$? What are the inverses in such a group?

6.2.2 *Hierarchical Structure

A directed n-skeleton is automatically endowed with a natural hierarchy of vertices. Using binary labels for the vertices of an n-skeleton, a vertex is said to belong to the k-th hierarchy if its binary label contains exactly k zeros. Having the same number of zeros is obviously an equivalence relation between vertices. Thus, the hierarchies can be regarded as equivalence classes. The number of elements (vertices) in the k-th hierarchy is the combinatorial number $C_k^n = \frac{n!}{k!\,(n-k)!}$. Thus, for a given n, the number of vertices in each class can be read off from the n-th row of Pascal's triangle.

Exercise 6.5 Prove that an equivalent way to define the hierarchy to which a vertex belongs is by counting the number of edges adjacent from this vertex.

A pictorially suggestive way of thinking of the vertex hierarchies is obtained by rotating the coordinate system so that the origin stands at the top and the last node (with a binary representation of all ones) lies at the bottom, like a spinning top. For $n = 2$ and $n = 3$ the corresponding pictures are demonstrated in Fig. 6.7.

For higher dimension $n > 3$ these perspective-inspired representations lose their immediacy, whereby the higher hierarchies can be placed above their lower counterparts while still preserving the intuitive perception of the square or the cube. Nevertheless, sacrificing this intuition, the graph plotted in the plane is still a graph topologically equivalent to the corresponding directed n-skeleton. This kind of representation is displayed in Fig. 6.8 for $n = 4$. The tesseract is not immediately intuited, but it is still there in terms of the relations between vertices and edges.

Exercise 6.6 Compare Fig. 6.8 with Fig. 6.3 to ascertain that they are topologically equivalent, namely, all the relations of adjacency between vertices are preserved.

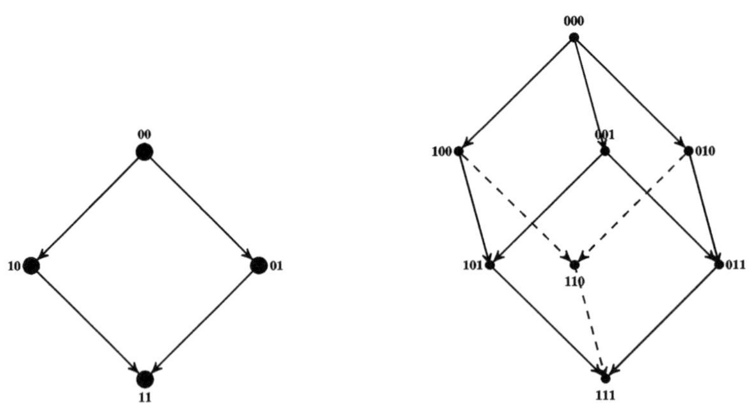

Fig. 6.7 Hierarchy of vertices for $n = 2$ (left), and $n = 3$ (right). Notice the 1-2-1 and 1-3-3-1 patterns of the hierarchies

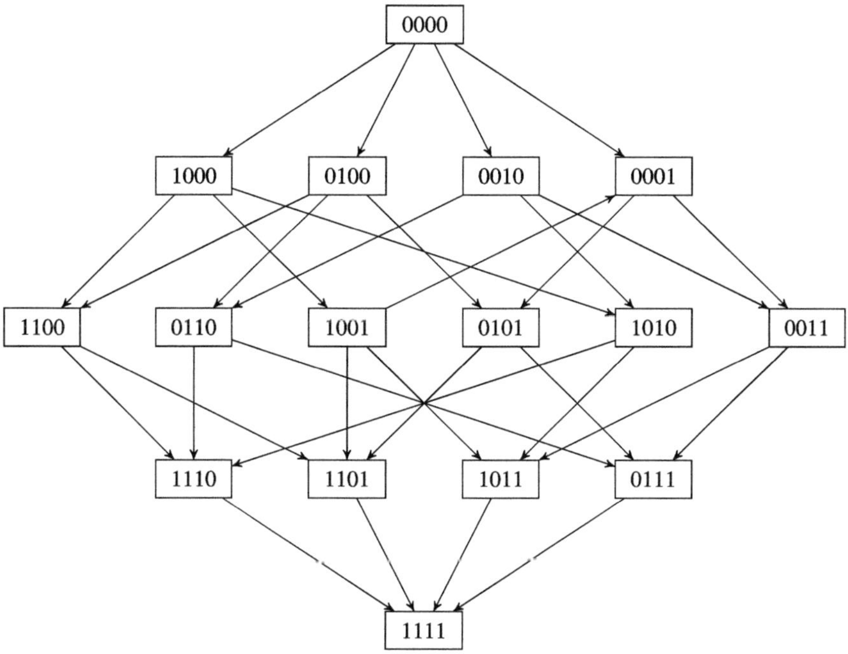

Fig. 6.8 Hierarchical structure of the vertices for $n = 4$. Notice the 1-4-6-4-1 pattern of the various hierarchies

Exercise 6.7 Regardless of the paraphernalia of the theory of groupoids, consider a mixture of n constituents. It is a self-evident truth that such a mixture can also be regarded as a mixture of sub-mixtures of m constituents, with $m = 1, \ldots, n-1$. For example, a ternary mixture ($n = 3$) can be conceived as a mixture containing single substances ($m = 1$), and binary mixtures ($m = 2$). Let us designate the number of constituents of each sub-mixture as its *rank*, varying from $m = 0$ (the empty mixture) up to $m = n$ (the total mixture). Show that, if we construct a directed graph where the nodes represent sub-mixtures and where each edge is an arrow pointing from one sub-mixture towards another in the immediately higher rank (of which the former is a constituent of the latter), this graph is the same as the graph of an oriented n-skeleton.

Chapter 7
n-fold Groupoids

Abstract An n-fold groupoid is a natural generalization of the idea of a double groupoid. Starting from n simple groupoids with a common base of objects, it is possible to proceed inductively to arrive at the notion of an n-fold groupoid as a set Q with n different groupoid structures, the base of each of which is an $(n-1)$-fold groupoid. The elements of Q can be regarded as hypercube skeletons. As expected, it is possible to define the material n-fold groupoid associated with a composite or a metamaterial of n constituents. Interestingly, the natural hierarchical structure of an n-skeleton can be interpreted physically in a composite in terms of the various sub-mixtures made up of all possible combinations of some of the constituents.

7.1 Objective n-Skeletons

7.1.1 Introduction

We have by now acquired considerable familiarity with the concepts of simple (that is, onefold) and double (or twofold) groupoids. The line of thought adopted in Chap. 4 to introduce double groupoids can be summarized as follows:

1. The point of departure is the presupposed existence of two simple groupoids, $\mathcal{Z}_1 \rightrightarrows \mathcal{B}$ and $\mathcal{Z}_2 \rightrightarrows \mathcal{B}$. The fact that these simple groupoids have a common set of objects $\mathcal{B}$ is what suggests the creation of a new entity, a double groupoid Q, that encompasses and synthesizes the possible interactions of the two constituent (or side) groupoids.
2. The new entity Q is itself endowed with two groupoid structures whose sets of objects are, respectively, $\mathcal{Z}_2$ and $\mathcal{Z}_1$, namely, the arrows of the second and the first side groupoid. Accordingly, using a somewhat metaphorical language, we referred to the arrows of each of the two groupoid structures of Q as hyper-arrows. For example, a hyper-arrow of the first structure is an entity that points from an arrow of $\mathcal{Z}_2$ (the source) to another arrow of $\mathcal{Z}_2$ (the target).
3. We found eventually that a convenient way to entrench the expected properties of Q was to introduce the concept of an ideal cell in the shape of the standard unit

© The Author(s), under exclusive license to Springer Nature Singapore Pte Ltd. 2026
M. Epstein, *An Elementary Introduction to* n*-fold Groupoids*,
Surveys and Tutorials in the Applied Mathematical Sciences 19,
https://doi.org/10.1007/978-981-95-1869-2_7

square cell $\mathcal{S}_2$ with oriented edges, shown in Fig. 4.1. The vertices of $\mathcal{S}_2$ are ideal geometric points (say, V_0, V_1, V_2, V_3) in the plane. The Cartesian coordinates, (X_1, X_2), of these points are the four distinct ordered pairs that can be formed with the binary digits 0 and 1.

4. The elements of $\mathcal{Q}$ were obtained by assigning to each of the ideal vertices V_i an object $W_i \in \mathcal{B}$. Moreover, to each of the four oriented sides of $\mathcal{S}_2$ we assigned an arrow in one of the side groupoids. More specifically, to each of the two edges parallel to X_1 (namely $\overrightarrow{V_0V_2}$ and $\overrightarrow{V_1V_3}$) we assign, respectively, arrows (if any) of the groupoid $\mathcal{Z}_1$ with source W_0 and target W_2, and with source W_1 and target W_3. A similar assignment, but with arrows in $\mathcal{Z}_2$, is made for the two edges parallel to X_2. Figure 4.2 can be used as a reminder. Thus defined, the elements of $\mathcal{Q}$ can be conceived as *objective squares*.
5. Taking advantage of the operations of composition in the simple groupoids, it is not difficult to define in a natural way operations of composition of objective squares, in each groupoid structure of $\mathcal{Q}$, and to verify a condition of compatibility between these operations.

In this final chapter we intend to follow a similar approach to define n-fold groupoids. Not surprisingly, the geometric and graph-theoretic background presented in Chap. 6 will play a role in the intended process of generalization. Finally, we will also generalize the concept of material double groupoid, developed in Chap. 5, and contextualize it within a terminology suitable to the description of multi-constituent composites.

7.1.2 *Vertices*

Let $\mathcal{B}$ denote a set of objects and let $n > 0$ be a natural number. We choose a *tuple* $\mathcal{W}$ of 2^n points of $\mathcal{B}$ which we denote by $\mathcal{W} = \{W_0, W_1, \ldots, W_{2^n-1}\}$. Two or more points of $\mathcal{W}$ may not necessarily be distinct as objects of $\mathcal{B}$, but they are considered as different entries in $\mathcal{W}$. Since the unit n-cube has exactly 2^n vertices, it makes sense to define a bijection between $\mathcal{W}$ and the collection of vertices of S_n, the skeleton of the standard unit hypercube Q_n. For this purpose, we need to adopt a definite ordering of these vertices.

As explained in Sect. 6.1, a convenient way to introduce a natural ordering of the vertices of S_n is obtained by exploiting the fact that the coordinates of each vertex of the unit n-cube consist of an ordered sequence of zeros and ones. But this is precisely the binary representation of the first 2^n non-negative integers! Denoting the ordered set of vertices of S_n thus obtained by $\mathcal{V} = \{V_0, V_1, \ldots, V_{2^n-1}\}$, with indices as exemplified in Fig. 7.1, we obtain the desired bijection $f : \mathcal{W} \to \mathcal{V}$ by setting $V_i = f(W_i)$. Exploiting this natural bijection the various relations of adjacency can be pulled back to the tuple $\mathcal{W}$.

Exercise 7.1 For the tesseract sketched in Fig. 6.3 provide the (decimal) numbering of the nodes, as was done for the cube in Fig. 7.1. Can you discern a common pattern?

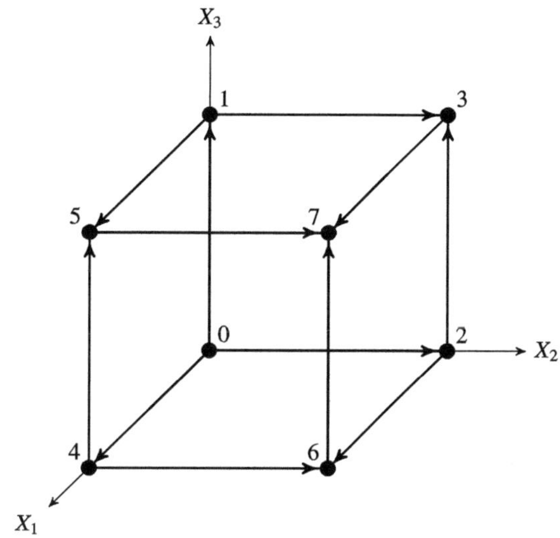

Fig. 7.1 Numbering the vertices of S_3 based on the binary system

7.1.3 Edges

Introducing further structure, let us suppose that we have been given (on physical or mathematical grounds) n simple (that is, ordinary) groupoids $\mathcal{Z}_I \rightrightarrows \mathcal{B}$ $(I = 1, \ldots, n)$ with common base $\mathcal{B}$. So far, we have introduced a bijection that allows us to grant a kind of materiality, or *objectivity* as it were, to the vertices of the ideal n-skeleton S_n. What we want to achieve now is a similar task of materialization *for the edges* of S_n. Since we have at our disposal n simple groupoids, and since the edges of S_n fall nicely into n different classes according to the coordinate axis to which they are parallel, as explained in Sect. 6.1, this task can be accomplished in a most natural way: each X_I edge of S_n (with tail V_i and head V_j) will be assigned an arrow (with source W_i and target W_j) in the groupoid $\mathcal{Z}_I$.

We denote by $\mathcal{P}^I_{XY}$ the set of arrows of $\mathcal{Z}_I$ with source at $X \in \mathcal{B}$ and target at $Y \in \mathcal{B}$.[1] The inherited relation of 'adjacency from' implies that each of the $2^n - 1$ entries $W_1, \ldots, W_{2^n-1}$ of $\mathcal{W}$ is adjacent from at least one other element of $\mathcal{W}$ (including W_0). For each ordered pair (W_a, W_b) of entries in $\mathcal{W}$ such that W_b is adjacent from W_a we choose an element of $P^I_{W_a W_b}$, where I is determined by the coordinate X_I in which $f(W_a)$ differs from $f(W_b)$. After all such pairs have been considered[2] we obtain (with a blatant abuse of notation) an injective map $g : S_n \to \mathcal{W} \times \mathcal{P}_{\mathcal{W}\times\mathcal{W}}$. We call the image of this map (if not empty) an *objective n-skeleton* T_n associated with $\mathcal{W}$. If each of the groupoids $\mathcal{Z}_I \rightrightarrows \mathcal{B}$ $(I = 1, \ldots, n)$ is transitive, at least

[1] Unless the groupoid $\mathcal{Z}_I \rightrightarrows \mathcal{B}$ happens to be transitive, for some pairs (X, Y) the set $\mathcal{P}^i_{XY}$ will be empty.

[2] If $P^I_{W_a W_b}$ is empty this construction is halted.

one objective n-skeleton exists for each 2^n-tuple $\mathcal{W}$. By having assigned to each objective oriented edge an arrow in a groupoid, we may regard the objective n-skeleton T_n as a kind of *weighted directed graph*, where the weights are arrows in various groupoids.

Exercise 7.2 Verify that for the case $n = 2$ this construction delivers the squares of a double groupoid, such as shown in Eq. (4.3). Notice how the edges parallel to the coordinate axis X_I $(I = 1, 2)$ are assigned arrows of the I–th groupoid. [Hint: refer back to Fig. 4.2.]

7.1.4 Projections

By virtue of its very definition, every objective n-skeleton T_n has objective h-faces for each $0 \leq h < n$. In turn, each of these h-faces is an objective h-skeleton. In particular, for $h = n - 1$ we obtain exactly $2n$ objective facets. As we learned in Sect. 6.1, these facets can be partitioned into n classes, each containing a pair of parallel X_I facets (for each $I = 1, \ldots, n$). Moreover, as we gather from Chap. 6, the objective n-skeleton T_n can be constructed, for any chosen $I = 1, \ldots, n$, from the two objective X_I-facets by joining their corresponding vertices with arrows of the simple groupoid $\mathcal{Z}_I$. A mental picture that one can form about these facts is suggested in Fig. 7.2 (omitting many of the details exhibited in the examples of Figs. 6.1, 6.2, 6.3 and 6.4). We may say, therefore, that an objective n-skeleton is naturally endowed with n source projections and n target projections, one of each for each $I = 1, \ldots, N$. This natural property of individual objective n-skeletons can be extended to the total collection, as we argue in what follows.

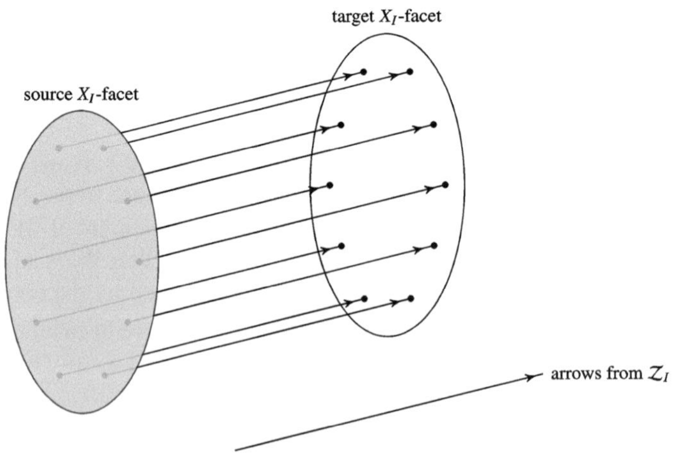

Fig. 7.2 Schematic generation of T_n from two objective facets

Let Q denote the set of all objective n-skeletons arising from n simple groupoids $\mathcal{Z}_I \rightrightarrows \mathcal{B}$ $(I = 1, \ldots, n)$ with a common base $\mathcal{B}$ of objects. Let, moreover, $Q_{\neg I}$ denote the set of all objective $(n-1)$-skeletons arising from the $n-1$ simple groupoids $\mathcal{Z}_I \rightrightarrows \mathcal{B}$ excluding the I-th groupoid. Each element of Q can be regarded as two (not necessarily distinct) elements of $Q_{\neg I}$ joined at their corresponding vertices by 2^{n-1} arrows of $\mathcal{Z}_I \rightrightarrows \mathcal{B}$. Accordingly, we can define two projection maps

$$\tilde{\alpha}_I : Q \to Q_{\neg I}, \qquad \tilde{\beta}_I : Q \to Q_{\neg I}. \tag{7.1}$$

These two maps, called respectively the I-th source map and the I-th target map in Q, are defined by applying the source and target projections α_I and β_I of $\mathcal{Z}_I \rightrightarrows \mathcal{B}$ to each X_I-edge of an objective n-skeleton, which delivers the 2^{n-1} vertices of the source and target facets, respectively.

Exercise 7.3 Prove that for each $I = 1, \ldots, n$ the projection maps $\tilde{\alpha}_I$ and $\tilde{\beta}_I$ are surjective. [Hint: for any given objective $(n-1)$-skeleton, construct the objective n-skeleton obtained by using unit arrows in the appropriate groupoid. How are the binary numberings of both skeletons related to each other?]

7.1.5 Hierarchical Structure

The set Q of objective n-skeletons has two projections over each of the n sets $Q_{\neg I}$ $(I = 1, \ldots, n)$. But each of these n sets is itself a collection of objective $(n-1)$-skeletons. Consequently, after this first step $(k = 1)$ we obtain further projections on sets of objective $(n-2)$-skeletons, which we may denote as $Q_{\neg IJ} = Q_{\neg JI}$. In other words, for the second step $(k = 2)$ we negate the appearance of two indices, I and J. This process can be pursued for exactly n steps. We remark that for the penultimate step $(k = n-1)$, when all indices are negated except for one, say the index I, we recover the simple groupoid $\mathcal{Z}_I$. When all indices are negated $(k = n)$, we recover the base set $\mathcal{B}$. Figure 7.3 shows all the projection pairs for the case $n = 4$.

To calculate the total number of projection pairs involved in this hierarchy of projections, we observe that before the k-th iteration step exactly $k-1$ distinct subscripts have been suppressed. Consequently, the number of sets of $(n-(k-1))$-hypercubes before this step is given by the combinatorial number

$$C^n_{n-(k-1)} = \frac{n!}{(k-1)!\,(n-(k-1))!}. \tag{7.2}$$

At the same time, each of these hypercubes has $n-(k-1)$ classes of opposite facets over which to project. We conclude that the total number of projection pairs (source

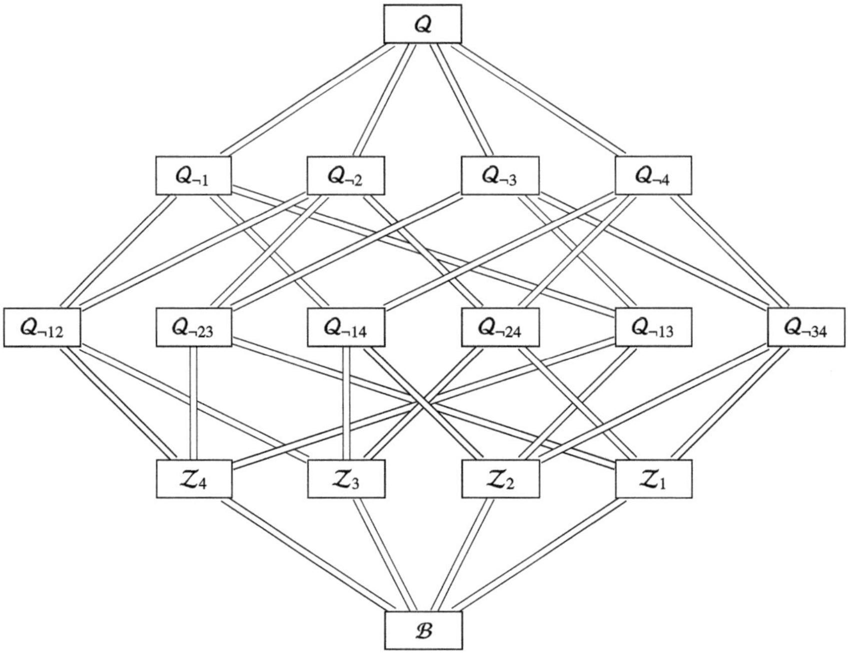

Fig. 7.3 Hierarchical structure of the projection maps for $n = 4$

and target) in n-dimensions is obtained as

$$P_n = \sum_{k=1}^{n} (n - (k-1)) \frac{n!}{(k-1)!(n-(k-1))!} = n \sum_{j=0}^{n-1} \frac{(n-1)!}{j!(n-1-j)!} = n\, 2^{n-1}. \tag{7.3}$$

Exercise 7.4 Carefully explain each step of the derivation of Eq. (7.3), or propose a shorter procedure (given the simplicity of the final result). Notice that this result happens to be the same as the number of edges of the n-hypercube.

Exercise 7.5 Check that for the cases $n = 2, 3, 4$ the formula (7.3) delivers the results that would be obtained by counting double arrows in a graph such as that of Fig. 7.3.

Proposition 7.1 *The hierarchical structure of the pairs of projections of the set $\mathcal{Q}$ of objective n-skeletons has the (combinatorial) topology of the standard directed n-skeleton $\mathcal{S}_n$.*

The proof of this result is obtained by comparing the hierarchical structure of the projections with the hierarchical structure of the vertices of $\mathcal{S}_n$ pursued in Sect. 6.2.2.

Exercise 7.6 Carry out the comparison just alluded to. Notice the identical topology of Figs. 6.8 and 7.3.

7.1.6 Composition of Objective n-Skeletons

Let T_n and T'_n be two n-skeletons in $\tilde{\mathcal{Z}}$ satisfying the condition

$$\tilde{\beta}_I(T'_n) = \tilde{\alpha}_I(T_n). \tag{7.4}$$

Because of the way in which these projections were defined (in terms of the projections α_I and β_I), the corresponding X_I edges of T_n and T'_n automatically satisfy the tip-to-tail condition as arrows of the I-th simple groupoid $\mathcal{Z}_I \rightrightarrows \mathcal{B}$ and they can be composed accordingly as such. We define the composition $T_n \odot_I T'_n$ as the objective n-skeleton such that

$$\tilde{\alpha}_I(T_n \odot_I T'_n) = \tilde{\alpha}_I(T'_n), \tag{7.5}$$

$$\tilde{\beta}_I(T_n \odot_I T'_n) = \tilde{\beta}_I(T_n), \tag{7.6}$$

and whose X_I-edges are obtained as the composition in the simple groupoid Z_I of the corresponding X_I edges of the factors.

Example 7.1 (The Case $n = 3$) Two objective 3-skeletons, as depicted in Fig. 7.4, are to be composed according to the composition operator $\odot_2$. Condition 7.4 prescribes that for this composition to be possible we must have

$$\hat{a}_1 = a_2, \qquad \hat{a}_3 = a_4, \qquad \hat{c}_1 = c_2, \qquad \hat{c}_3 = c_4. \tag{7.7}$$

The composition affects only the edges of class X_2, as indicated at the lower part of Fig. 7.4. Conditions (7.5) and (7.6) are fulfilled.

□

Following along these lines, we can define the identities associated with the composition $\odot_I$ by identifying with each other the two X_I-facets of T_n, which we have denoted by $T^{I,0}_{n-1}$ and $T^{I,1}_{n-1}$, and adopting for each of the linking X_I-edges the identity element in the corresponding simple groupoid. With this composition law and the corresponding identities, we can endow the set $\tilde{\mathcal{Z}}$ with the structure of a simple groupoid over the set $\tilde{\mathcal{Z}}_{\bar{I}}$. Letting I range from 1 to n we obtain a collection of n groupoid structures, one for each of the set of 'objects' $\tilde{\mathcal{Z}}_{\bar{I}}$. The 'arrows' of each of these simple groupoids consist of the set of 2^{n-1} oriented X_I-edges of an n-skeleton.

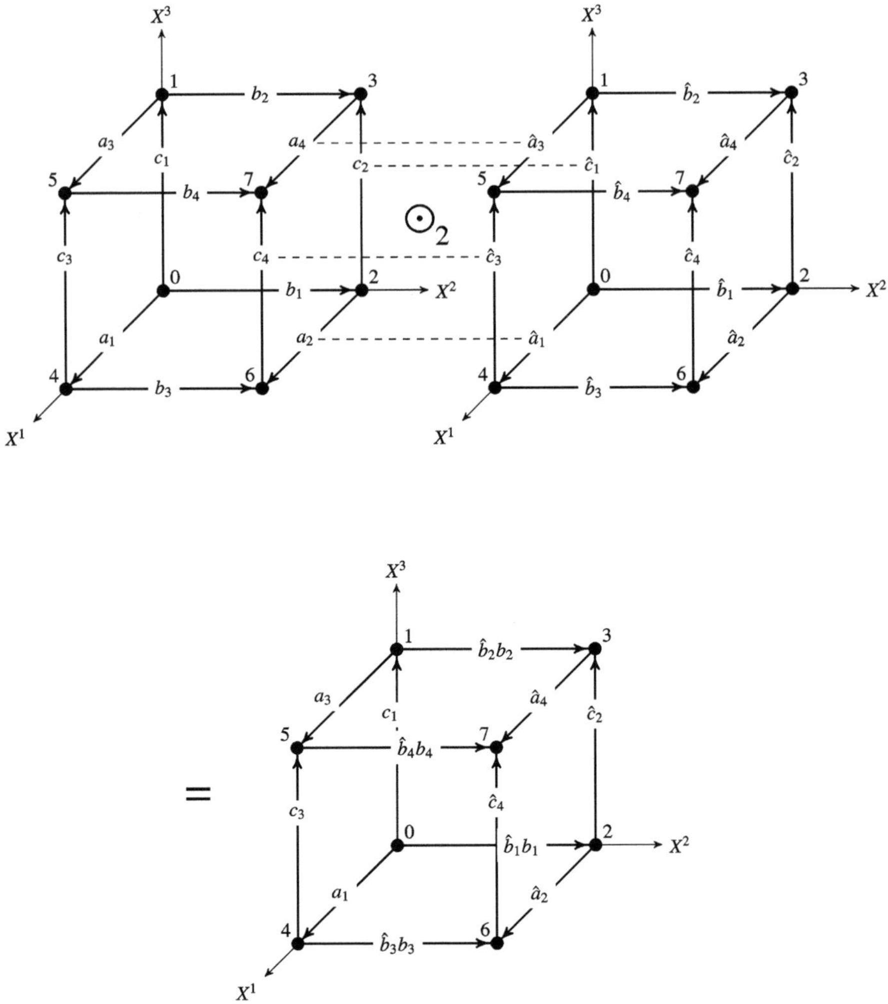

Fig. 7.4 Composition of two objective 3-skeletons in the X^2 direction

7.1.7 Coarse n-fold Groupoids

Can these n simple groupoid structures be unified under the umbrella of an n-fold groupoid? To obtain a positive answer to this question we need to show that the various compositions satisfy the compatibility conditions

$$(T_n \odot_I T_n') \odot_J (T_n'' \odot_I T_n''') = (T_n \odot_J T_n'') \odot_I (T_n' \odot_J T_n''') \tag{7.8}$$

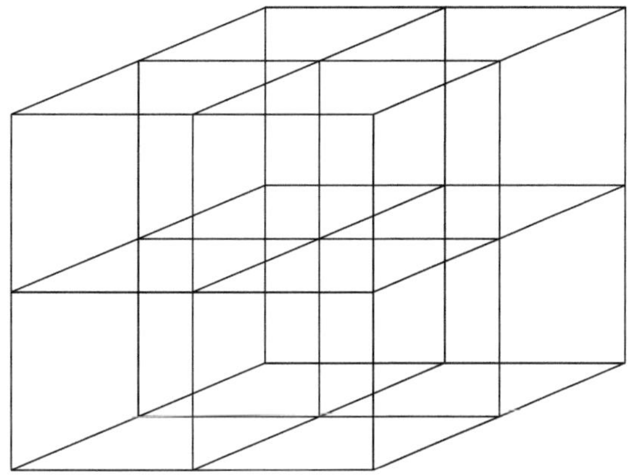

Fig. 7.5 Compatibility argument for $n = 3$

for each distinct pair I, J whenever the operations are possible. The proof is straightforward, since each composition affects only the edges parallel to one direction. We will refer to the set $\tilde{\mathcal{Z}}$ and its n groupoid structures over $\tilde{\mathcal{Z}}_{\bar{I}}$ as an n-fold groupoid. The units and inverses of each $\tilde{\mathcal{Z}}_I$ are clearly defined by rendering the corresponding X_I-edges as units and inverses in the ordinary groupoid $\mathcal{Z}_I \rightrightarrows \mathcal{B}$.

Exercise 7.7 For the case $n = 3$ verify that the compatibility condition (7.8) holds true. Figure 7.5 can help to advance the argument.

At this point it seems appropriate to revive our premonitory definition of an n-fold groupoid anticipated in Sect. 4.2.1. The more precise recursive definition reads:

Definition 7.1 An n-fold groupoid arising from n simple groupoids with common base $\mathcal{B}$ is a set $\mathcal{Q}$ equipped with n groupoid structures each of which has as a base an $(n-1)$-fold groupoid arising from one of the n different combinations of $n-1$ of the n simple groupoids of departure. Denoting the compositions of these groupoid structures of $\mathcal{Q}$ by $\odot_I$ $(I = 1, \ldots n)$, the following *compatibility condition* must be satisfied for every pair of indices $I \neq J$:

$$(\mathbb{A}\odot_I\mathbb{B})\odot_J(\mathbb{C}\odot_I\mathbb{D}) = (\mathbb{A}\odot_J\mathbb{C})\odot_I(\mathbb{B}\odot_J\mathbb{D}), \qquad \forall\, \mathbb{A}, \mathbb{B}, \mathbb{C}, \mathbb{D} \in \mathcal{Q}, \tag{7.9}$$

whenever the operations are possible.

Given n simple groupoids, the corresponding *coarse n-fold groupoid* is obtained by considering all the possible objective n-skeletons that can be obtained by assigning the X_I-edges to arrows of the simple groupoid $\mathcal{Z}_I \rightrightarrows \mathcal{B}$. The hierarchical structure of the projection maps can now be understood as the hierarchical structure of all the m-fold groupoids involved, where $m = 1, \ldots, n$. In other words, we can formulate the following corollary of Proposition 7.1

Corollary 7.1 *The number of m-fold groupoids of all orders* $(m = 1, \ldots, n)$ *involved in an n-fold groupoid is equal to the number of vertices of the standard directed n-skeleton* $\mathcal{S}_n$. *Moreover, the (combinatorial) topology of the hierarchical* *structure of the interrelation between these groupoids is the same as that of the vertices of* $\mathcal{S}_n$.

This is a remarkable property of self-replication of a hierarchical structure.

Remark 7.1 The standard diagram (4.1) used to represent the four groupoids involved in a double groupoid is also often represented as in (7.10), which is more suggestive of the hierarchies in this simplest instance of an n-fold groupoid.

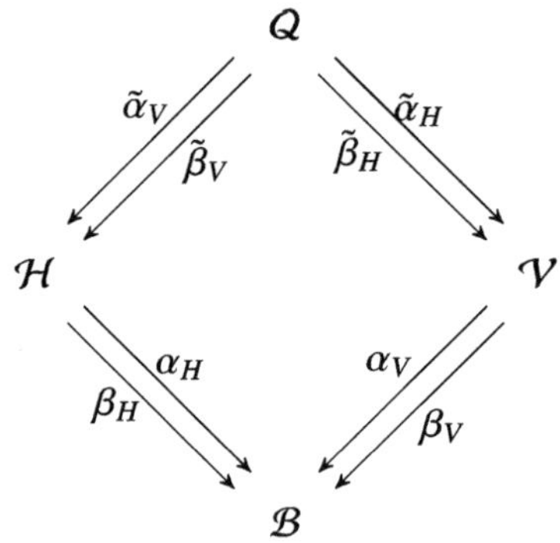

(7.10)

Either way, one can hardly miss the detail that the shape of this diagram is a square and, coincidentally, the elements in the double groupoid are squares. What Corollary 7.1 asserts is that this apparent coincidence can be extended to all n-fold groupoids. For the case $n = 3$, for example, we obtain the diagram (7.11) below, which can be viewed as the perspective representation of a cube.

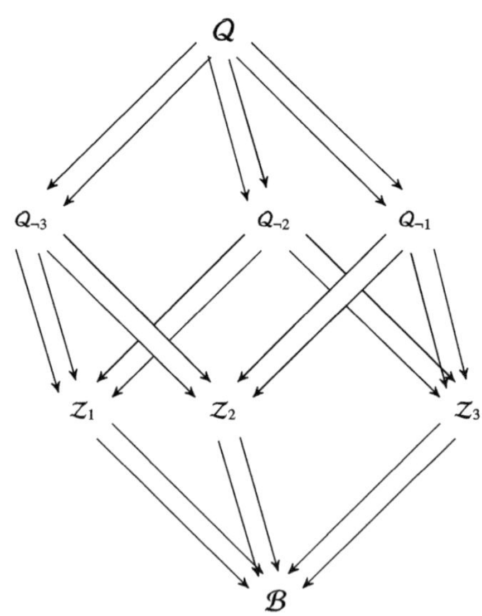

(7.11)

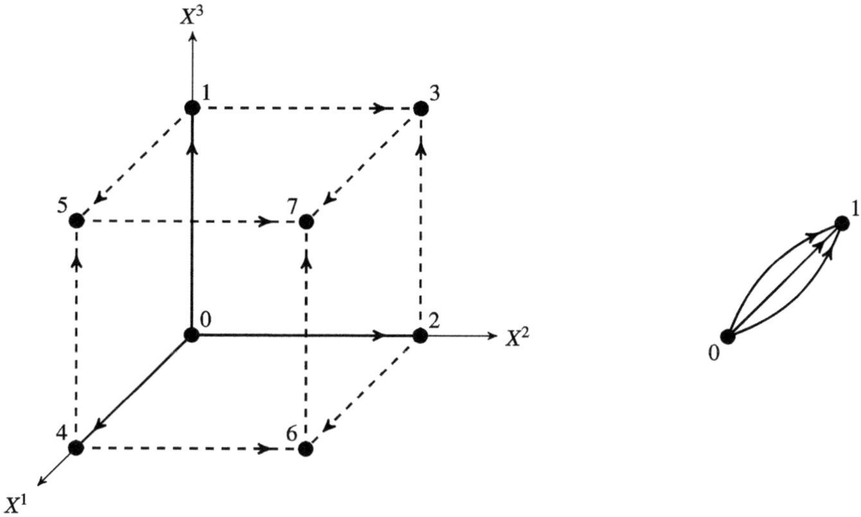

Fig. 7.6 An element in the core, where identity *arrows* are *dashed*

Considering the recursive character of our definition of an n-fold groupoid, it should be clear that every 2-face of an n-fold groupoid is a double groupoid in the usual sense of the term [2, 6, 9].

Exercise 7.8 Identifying the onefold groupoid with a simple groupoid (as defined in Chap. 1), apply the recursive definition 7.1 to obtain a twofold groupoid. Compare the result with that of a double groupoid as defined in Chap. 4.

To obtain the *core groupoid* of an n-fold groupoid we set all the edges, except those adjacent from the vertex W_0, of each objective n-skeleton to the identity element of the corresponding groupoid. Effectively, therefore all the vertices but W_0 collapse into a single vertex, W_1 say, and we obtain a onefold groupoid whose arrows are a bundle of n arrows, one from each groupoid. This is schematically represented for the case $n = 3$ in Fig. 7.6, which is intended only to be suggestive of the abstract situation.

7.2 Material n-fold Groupoids

In Sect. 7.1.7 we have introduced the *coarse* n-fold groupoid of all possible n-skeletons constructed from the arrows of n given simple groupoids. Imposing further restrictions on the groupoids of departure and on the selection of their arrows we can obtain smaller n-fold groupoids with a precise physical meaning, such as is the case in the theory of multi-composites and mixtures [10, 11].

Consider the special case in which $\mathcal{B}$ is a material body and each of the groupoids $\mathcal{Z}_I \rightrightarrows \mathcal{B}$ ($I = 1, \ldots, n$) is a subgroupoid of $\Pi^1(\mathcal{B}, \mathcal{B})$, the 1-jet groupoid of $\mathcal{B}$, whose arrows are arbitrary elements of $GL(3, \mathbb{R})$. In this case, the weights of the edges of our objective skeletons (which can be regarded as weighted directed graphs, as explained in Sects. 6.2.1 and 7.1.3) become, in any given chart of $\mathcal{B}$, elements of the general linear group $GL(3, \mathbb{R})$.

Because of the above stated assumptions, the weights of two objective edges of different classes can be composed provided the target of an incoming edge is the same as the source of an edge issuing from the same objective vertex. Using the language of graph theory, a *walk* is a sequence of successively adjacent edges. In our weighted directed graph the weight of an edge traversed against its orientation is taken as the inverse of the original weight. The total *length* of the walk is the composition of the weights of the traversed edges. A *closed walk* is a walk in which the first and last vertices coincide.

Theorem 7.1 *The total weight of every closed walk in $T_n \in Q$ is the group identity if, and only if, every 2-face is commutative.*

We will refer to such a *conservative* n-fold groupoid as a *material n-fold groupoid*. The 'only if' part of the theorem is trivial. The proof of the 'if' part can be obtained by induction as follows. The theorem is trivially true for $n = 2$. Assume it to be true for some value $n - 1$. Given a closed walk in T_n:

1. If the walk is completely contained in an $(n - 1)$-face of T_n the theorem is obviously true.
2. If not, consider the partition of T_n into two X_I-facets. Since in a closed walk the initial and terminal vertices coincide, any closed walk must comprise an even number of X_I edges, just like the number of bridge traversals between two shores connected by various bridges. Consider, therefore, a pair of successive traversals. If the connecting edge happens to be the same for both traversals, we conclude that their combined contribution to the total weight of the closed walk is the group identity, since the same edge is traversed in opposite directions.
3. Otherwise, if the connecting edges are different, let A and B denote the corresponding vertices in the two X_I-facets joined by these two edges. We can join A and B by an identical finite path in each of the facets. Consider a single edge AA' of these paths, as shown in Fig. 7.7. Together with the two edges, say E_1 and E_2 joining the corresponding vertices, they constitute a 2-face of T_n. By commutativity, therefore, the weight contributed by E_1 is identical to that contributed by E_2. Continuing along the path joining A with B, we can invoke the reasoning adduced in the previous item (where the two edges coincided).

Corollary 7.2 *A mixture of n uniform constituents is materially uniform if, and only if, the core of its material n-fold groupoid is transitive.*

If the mixture is not uniform we say that there are sources of misalignment, namely, a defectivity arising from an incompatibility between the constituents. We remark, however, that even if all the constituents are compatible with a fixed one,

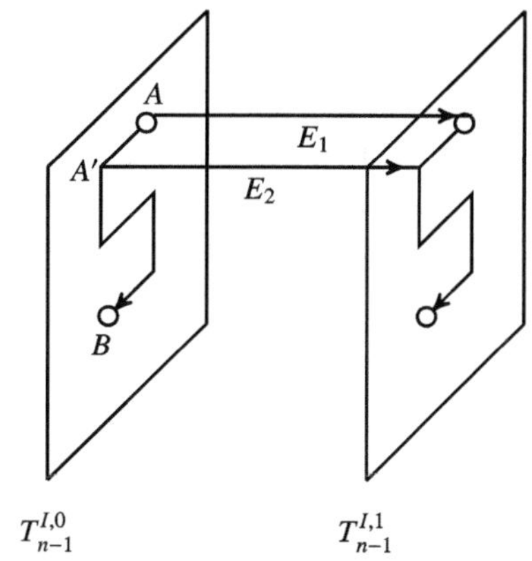

Fig. 7.7 Independence of weight for parallel traversing edges

the mixture is not necessarily uniform. In other words, in the case of continuous symmetry groups, the core of the material n-fold groupoid may fail to be transitive, due to the extra degrees of freedom afforded by the symmetries. If this observation is correct, we can assert that a different kind of defectivity has been identified beyond that provided in [11] for strictly binary mixtures.

Exercise 7.9 Prove that in a mixture of n constituents with the trivial symmetry group, if one of them is aligned with the remaining $n - 1$ constituents the mixture is necessarily uniform. Give a counterexample for the case in which at least one of the constituents is isotropic.

As an example, which may be applicable to metamaterials, we can consider a combination of three materials, one of which is an isotropic solid in a homogeneous configuration. The other two constituents have discrete symmetry groups and are in different states of contorted aeolotropy. This terminology is used to describe a stress-free state of a uniform material whereby all the material isomorphisms are pure rotations [29]. The isotropic constituent is, therefore, well aligned with each of the other two constituents, since isotropy affords a degree of freedom of rotation. If the composite is layered, as illustrated in Sect. 5.1, and the isotropic constituent is sandwiched between the other two constituents, we obtain an inert mixture. A similar situation arises in a spatial arrangement schematically conceived as a regular scaffold of cubic boxes whose walls are made of the isotropic constituent, as shown in Fig. 7.8, in which the boxes are alternately filled with one of the other materials (so that the latter are not in contact with each other). If, on the other hand, the scaffold is made of one of the anisotropic materials, those boxes filled with the other anisotropic constituent may react and evolve in time (perhaps due to the existence of a chemical potential associated with the misalignment). This kind of situation may arise in applications to cell and tissue engineering [14, 33]. More generally, an evolution in time of the overall groupoid structure can be induced artificially in *tunable* metamaterials [27, 35, 36].

Fig. 7.8 Alternating filling in metamaterial

Example 7.2 (Time Evolution) One or more of the constituents of a material n-fold groupoid may evolve in time due to internal causes (such as a biological process triggered by misalignment between two constituents), or external agents (such as the application of an electric field in tunable metamaterials). In Example 5.1 we considered a possibly deliberate distribution of defects of misalignment. Suppose, instead, that we are dealing with a biological composite and that, spurred by the very presence of those defects, the second (hatted) constituent becomes susceptible to remodel its implants in an effort to relax the discrepancy with the structure of the first constituent, which remains unaltered as a stronger stable substratum. Recall that the local measure of defectivity is the referential gradient $\mathbf{C}$ of the tensor $\mathbf{A}$, defined in Eq. (5.3). The components of this tensor are

$$A^{\alpha}_{\beta} = \left(\hat{P}^{-1}\right)^{\alpha}_{M} P^{M}_{\beta}. \tag{7.12}$$

In terms of components, therefore, the local measure $\mathbf{C}$ of misalignment is

$$C^{\alpha}_{\beta I} = \frac{\partial A^{\alpha}_{\beta}}{\partial X^{I}}. \tag{7.13}$$

We shall assume an *evolution constitutive law* of the form

$$\frac{\partial \hat{P}^{J}_{\alpha}}{\partial t} = \hat{P}^{J}_{\beta}\, g^{\beta}_{\alpha}\left(\mathbf{C}, \nabla\mathbf{C}\right), \tag{7.14}$$

where t is the time variable, $\mathbf{g}$ is a proposed material law of evolution, and ∇ indicates the referential gradient operator. For the purposes of our illustration, we may adopt a strictly linear dependence on the gradient of $\mathbf{C}$ alone, that is,

$$g^{\beta}_{\alpha} = D^{\beta\nu KL}_{\alpha\mu} \frac{\partial^{2} A^{\mu}_{\nu}}{\partial X^{K}\partial X^{L}}, \tag{7.15}$$

where $D_{\alpha\mu}^{\beta\nu JK}$ are constant coefficients. We may further simplify the problem by assuming that

$$D_{\alpha\mu}^{\beta\nu KL} = -\delta_{\alpha}^{\nu}\delta_{\mu}^{\beta}E^{KL}, \tag{7.16}$$

where $\mathbf{E}$ is a symmetric positive-definite material tensor to be determined experimentally. Introducing these constitutive assumptions into Eq. (7.14) we obtain the following system of second-order PDEs,

$$\frac{\partial \hat{P}_{\alpha}^{J}}{\partial t} = -E^{KL}\,\hat{P}_{\mu}^{J}\,\frac{\partial^2 A_{\alpha}^{\mu}}{\partial X^K \partial X^L}, \tag{7.17}$$

to be solved with appropriate initial and boundary conditions.

As an example, setting $\mathbf{E}$ to the unit 2×2 matrix, using the initial conditions

$$[\hat{P}](0) = \begin{bmatrix} \cos\theta(x, y, 0) & -\sin\theta(x, y, 0) \\ \sin\theta(x, y, 0) & \cos\theta(x, y, 0) \end{bmatrix}, \tag{7.18}$$

with

$$\theta(x, y, 0) = 0.5\sin^2(\pi\, x/2) + 0.3\sin^2(\pi\, y), \tag{7.19}$$

and imposing Neumann (zero-flux) boundary conditions on a square with side 2, the solution (in terms of streamlines of the implants, obtained with Mathematica®) is shown in Fig. 7.9 for times $t = 0, 0.02, 0.20$. It can be observed that at the final time the misalignment has disappeared. We remark that the implants $\hat{\mathbf{P}}$ have converged to a constant field, implying that the field $\mathbf{A}$ has also become constant, even if $\hat{\mathbf{P}} \neq P$.

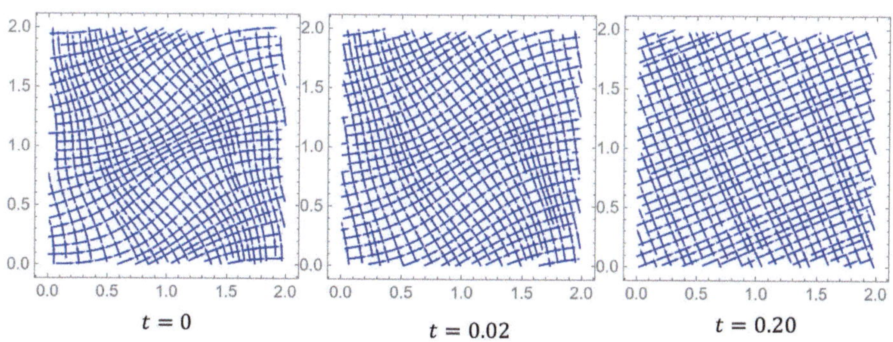

Fig. 7.9 Time evolution of a material groupoid in a composite

Remark 7.2 Noting that the governing equations are diffusive, any possible inconsistency between initial and boundary conditions is eliminated almost immediately by dissipation. It is also noteworthy that the orthogonality of the initial frame (if not necessarily the length of its vectors) is preserved by the solution as a consequence of the assumed form of g_{α}^{β}.

Appendix A
Solutions of Selected Exercises

A.1 Chapter 1

Exercise 1.1 Recall that in this group the symbol r_i represents a counterclockwise rotation of $(72^o)\,i$. Moreover, a rotation of 360° has been identified with the group unit r_0, and also $r_i\, r_j = r_{i+j}$. Therefore,

$$r_1\, r_4 = r_2\, r_3 = r_3\, r_2 = r_4\, r_1 = r_0. \tag{A.1}$$

Exercise 1.2 For example,

$$f_0(r_1(a_0)) = f_0(a_1) = a_4, \tag{A.2}$$

and

$$r_1(f_0(a_0)) = r_1(a_0) = a_1. \tag{A.3}$$

More generally, we can represent the effects of the products $f_0 r_1$ and $r_1 f_0$ on the whole star, as shown in Figs. A.1 and A.2, respectively and conclude that

$$f_0 r_1 = f_2, \tag{A.4}$$

and

$$r_1 f_0 = f_3, \tag{A.5}$$

In this way we can obtain the complete *multiplication table* of this group, as shown in Table A.1. The groups of symmetries of regular polygons are known as *dihedral groups*.

© The Author(s), under exclusive license to Springer Nature Singapore Pte Ltd. 2026
M. Epstein, *An Elementary Introduction to* n*-fold Groupoids*,
Surveys and Tutorials in the Applied Mathematical Sciences 19,
https://doi.org/10.1007/978-981-95-1869-2

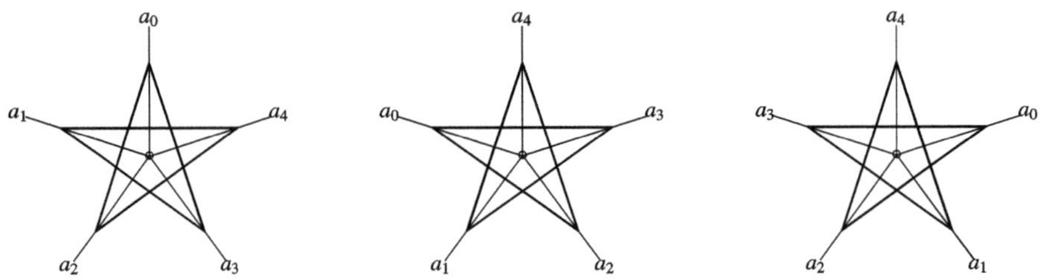

Fig. A.1 First apply r_1 and then f_0; the result is f_2

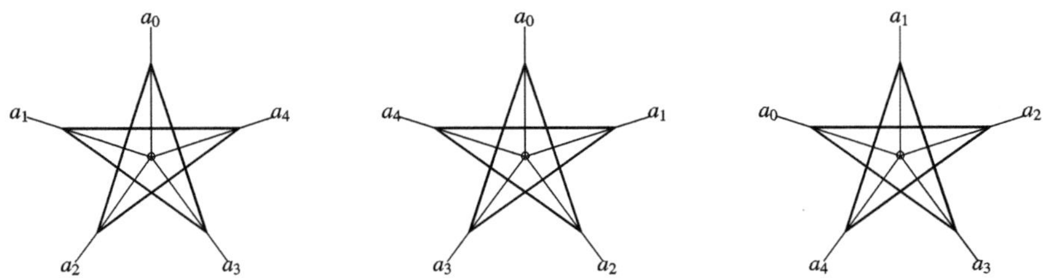

Fig. A.2 First apply f_0 and then r_1; the result is f_3

Table A.1 Group multiplication table

	r_0	r_1	r_2	r_3	r_4	f_0	f_1	f_2	f_3	f_4
r_0	r_0	r_1	r_2	r_3	r_4	f_0	f_1	f_2	f_3	f_4
r_1	r_1	r_2	r_3	r_4	r_0	f_3	f_4	f_0	f_1	f_2
r_2	r_2	r_3	r_4	r_0	r_1	f_1	f_2	f_3	f_4	f_0
r_3	r_3	r_4	r_0	r_1	r_2	f_4	f_0	f_1	f_2	f_3
r_4	r_4	r_0	r_1	r_2	r_3	f_2	f_3	f_4	f_0	f_1
f_0	f_0	f_2	f_4	f_1	f_3	r_0	r_3	r_1	r_4	r_2
f_1	f_1	f_3	f_0	f_2	f_4	r_2	r_0	r_3	r_1	r_4
f_2	f_2	f_4	f_1	f_3	f_0	r_4	r_2	r_0	r_3	r_1
f_3	f_3	f_0	f_2	f_4	f_1	r_1	r_4	r_2	r_0	r_3
f_4	f_4	f_1	f_3	f_0	f_2	r_3	r_1	r_4	r_2	r_0

Exercise 1.5 Assume two unit elements, e_1 and e_2. Since a unit (according to the definition of a group) can be composed both on the left and on the right, we must have

$$e_2 = e_1\, e_2 = e_1, \tag{A.6}$$

which proves the uniqueness of the unit element. Similarly, assume that an element g has two inverses, a and b. Then

$$a = ea = (bg)a = b(ga) = be = b. \tag{A.7}$$

Exercise 1.6
(c) This is not a group since the product of two negative numbers is not a negative number.
Exercise 1.10

(a) Since the group of rotations of the five-pointed star is commutative, any conjugation reproduces the original element. As a result, the conjugacy classes are singletons.
(b) Using the multiplication table and Eq. (1.16), we obtain the following conjugacy classes:

$$\{r_0\}, \quad \{r_1, r_4\}, \quad \{r_2, r_3\}, \quad \{f_0, f_1, f_2, f_3, f_4\}. \tag{A.8}$$

Exercise 1.12 To verify that the conjugate $a\mathcal{H}a^{-1}$ of a subgroup $\mathcal{H} \subset \mathcal{G}$ is also a subgroup of $\mathcal{G}$ we need to prove that $a\mathcal{H}a^{-1}$ is closed under the operations of multiplication and inversion. Two generic elements of $a\mathcal{H}a^{-1}$ must be of the form ah_1a^{-1} and ah_2a^{-1}, for some $h_1, h_2 \in \mathcal{H}$. Therefore, by associativity in $\mathcal{G}$,

$$\left(ah_1a^{-1}\right)\left(ah_2a^{-1}\right) = a(h_1h_2)a^{-1}, \tag{A.9}$$

which proves that h_1h_2 is in $a\mathcal{H}a^{-1}$. Similarly, for each $h \in \mathcal{H}$,

$$\left(ah^{-1}a^{-1}\right)\left(aha^{-1}\right) = e, \tag{A.10}$$

which proves that the inverse of aha^{-1} is in $a\mathcal{H}a^{-1}$. More specifically, the inverse of an element in the conjugate is the conjugate of its inverse.
Exercise 1.17 Let v belong to $\mathcal{P}_{XY}$. The product $g = u^{-1}v$ is a loop-shaped arrow, namely, it belongs to $\mathcal{G}_X$. Therefore, v is of the form ug for some $g \in \mathcal{G}_X$. Conversely, let $w = ug'$ for some $g' \in \mathcal{G}_X$. By the rules of groupoid composition we conclude that w is a groupoid arrow with source at X and target at Y.

A.2 Chapter 2

Exercise 2.7 Consider a generic vector $\mathbf{v}$ in $\mathbb{R}^2$. It has a magnitude w and forms an angle α measured counter-clockwise from the x-axis. If we rotate this vector counter-clockwise by an angle θ, it will become a new vector $\mathbf{v}'$ with the same

Fig. A.3 Rotation of a vector in the plane

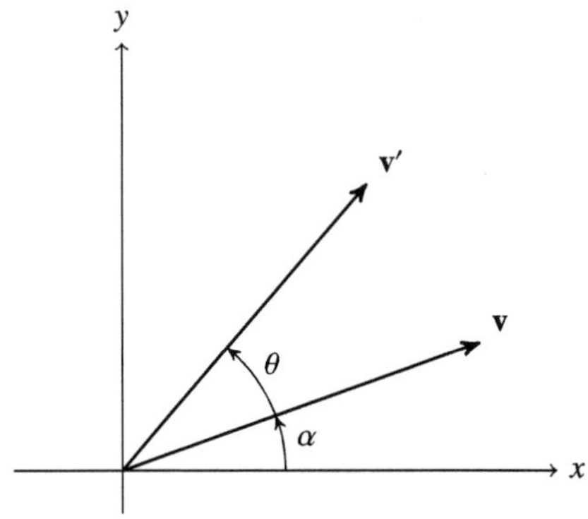

magnitude w and forming an angle of $\theta + \alpha$ with the x-axis, as shown in Fig. A.3. In components,

$$\{v'\} = w \begin{Bmatrix} \cos(\alpha + \theta) \\ \sin(\alpha + \theta) \end{Bmatrix} = w \begin{Bmatrix} \cos\alpha \, \cos\theta - \sin\alpha \, \sin\theta \\ \sin\alpha \, \cos\theta + \cos\alpha \, \sin\theta \end{Bmatrix} \tag{A.11}$$

$$= w \begin{bmatrix} \cos\theta & -\sin\theta \\ \sin\theta & \cos\theta \end{bmatrix} \begin{Bmatrix} \cos\alpha \\ \sin\alpha \end{Bmatrix} = \begin{bmatrix} \cos\theta & -\sin\theta \\ \sin\theta & \cos\theta \end{bmatrix} \{v\}.$$

In short, the general form of a rotation matrix is indeed

$$[R]_\theta = \begin{bmatrix} \cos\theta & -\sin\theta \\ \sin\theta & \cos\theta \end{bmatrix}. \tag{A.12}$$

To verify that all matrices of this form constitute a group, we check the multiplication of two such matrices and obtain

$$\begin{bmatrix} \cos\theta & -\sin\theta \\ \sin\theta & \cos\theta \end{bmatrix} \begin{bmatrix} \cos\phi & -\sin\phi \\ \sin\phi & \cos\phi \end{bmatrix} = \begin{bmatrix} \cos(\theta + \phi) & -\sin(\theta + \phi) \\ \sin(\theta + \phi) & \cos(\theta + \phi) \end{bmatrix}. \tag{A.13}$$

The unit matrix is recovered for $\theta = 0$. The inverse matrix is obtained by changing the sign of the angle of rotation, which gives the result

$$[R]_\theta^{-1} = [R]_{-\theta} = [R]_\theta^T. \tag{A.14}$$

The determinant of $[R]_\theta$ is equal to $+1$ for all values of θ, as follows from the basic trigonometric identity. In short, each of these matrices has the property that the inverse is equal to the transpose, and a determinant equal to $+1$. The converse is also true, as can be verified by enforcing these conditions. By extension, these matrices can be characterized as those that preserve the dot product of vectors and their relative orientation.

Exercise 2.9 For the case $n = 2$ we obtain

$$\begin{aligned}\sum_{i=1}^{2}\sum_{j=1}^{2}\epsilon_{ij} f_{1i} f_{2j} &= \cancel{\epsilon_{11} f_{11} f_{21}} + \epsilon_{12} f_{11} f_{12} + \epsilon_{21} f_{12} f_{21} + \cancel{\epsilon_{22} f_{12} f_{22}} \\ &= f_{11} f_{22} - f_{12} f_{21} = \det[F].\end{aligned} \tag{A.15}$$

For $n = 3$, omitting the vanishing terms with repeated indices, we obtain

$$\begin{aligned}\sum_{i=1}^{3}\sum_{j=1}^{3}\sum_{k=1}^{3}\epsilon_{ijk} f_{1i} f_{2j} f_{3k} &= \epsilon_{123} f_{11} f_{22} f_{33} + \epsilon_{132} f_{11} f_{23} f_{32} + \epsilon_{213} f_{12} f_{21} f_{33} \\ &\quad + \epsilon_{231} f_{12} f_{23} f_{31} + \epsilon_{312} f_{13} f_{21} f_{32} + \epsilon_{321} f_{13} f_{22} f_{31} \\ &= f_{11} f_{22} f_{33} - f_{11} f_{23} f_{32} - f_{12} f_{21} f_{33} \\ &\quad + f_{12} f_{23} f_{31} + f_{13} f_{21} f_{32} - f_{13} f_{22} f_{31} = \det[F].\end{aligned} \tag{A.16}$$

Exercise 2.13 Let $\{v\}$ be a column vector with components X, Y, Z. The equation of the unit sphere is

$$\{v\}^T \{v\} = 1. \tag{A.17}$$

Under the given linear transformation $[F]$ the vector gets transformed into the vector $\{v\}' = [F]\{v\}$. We conclude that the vector $\{v\}'$ components x, y, z satisfies the equation

$$\{v\}'^T \left([F][F]^T\right)^{-1} \{v\}' = 1. \tag{A.18}$$

We obtain the symmetric (and positive definite) matrix

$$[B] = [F][F]^T = \begin{bmatrix} 0.82 & 0.31 & 0.5 \\ 0.31 & 1.01 & -0.02 \\ 0.5 & -0.02 & 1.17 \end{bmatrix}. \tag{A.19}$$

Its inverse is

$$[B]^{-1} = \begin{bmatrix} 1.97698 & -0.623735 & -0.855523 \\ -0.623735 & 1.18722 & 0.286848 \\ -0.855523 & 0.286848 & 1.22521 \end{bmatrix}. \tag{A.20}$$

The equation of the ellipsoid is, therefore,

$$\langle x\ y\ z\rangle \begin{bmatrix} 1.97698 & -0.623735 & -0.855523 \\ -0.623735 & 1.18722 & 0.286848 \\ -0.855523 & 0.286848 & 1.22521 \end{bmatrix} \begin{Bmatrix} x \\ y \\ z \end{Bmatrix} = 1. \tag{A.21}$$

This equation is equivalent to the one given in the statement of the exercise, which was obtained by equating the magnitude of the vector $[F]^{-1}\{v\}'$ to 1. The quadratic form with the positive-definite symmetric matrix $[B]^{-1}$ can be used to deliver the corresponding positive eigenvalues which represent the reciprocals of the squares of the semi-axes. Accordingly (solving the characteristic cubic polynomial) these semi–axes are: 0.595598, 1.03336, and 1.25596. Their product must naturally be the same as the determinant of $[F]$, namely, 0.773 (which can be calculated directly from the given transformation matrix). Either way, the volume of the ellipsoid is then 77.3% of the volume of the original sphere.
Exercise 2.16 The counter-clockwise rotation by an angle ϕ about y is given by

$$[R]_y = \begin{bmatrix} \cos\phi & 0 & \sin\phi \\ 0 & 1 & 0 \\ -\sin\phi & 0 & \cos\phi \end{bmatrix}. \tag{A.22}$$

Similarly for a rotation of ψ about X we obtain

$$[R]_x = \begin{bmatrix} 1 & 0 & 0 \\ 0 & \cos\psi & \sin\psi \\ 0 & -\sin\psi & \cos\psi \end{bmatrix}. \tag{A.23}$$

The right-hand rule (with the thumb pointing in the positive direction of the axis of rotation) has been observed. For $\psi = \phi = 90^o$ we obtain

$$[R]_x[R]_z = \begin{bmatrix} 0 & -1 & 0 \\ 0 & 0 & -1 \\ 1 & 0 & 0 \end{bmatrix} \tag{A.24}$$

and

$$[R]_z[R]_x = \begin{bmatrix} 0 & 0 & 1 \\ 1 & 0 & 0 \\ 0 & 1 & 0 \end{bmatrix}. \tag{A.25}$$

The suggested experiment with a physical book or a brick should confirm these results. Start with the book upright with the pages parallel to the plane yz.
Exercise 2.18 [Warning: we are using here some of the terminology introduced in Chap. 3. Nevertheless, Eq. (A.30) below can be used to answer the exercise

independently.] The general idea is quite straightforward. Let $L_g : \mathcal{A} \to \mathcal{A}$ be a (left, say) action of a group $\mathcal{G}$ on a set $\mathcal{A}$. A *symmetry* of a subset $A \subset \mathcal{A}$ under the given action is an element $s \in \mathcal{G}$ such that

$$L_s(A) = A. \tag{A.26}$$

Let $f : \mathcal{A} \to \mathcal{B}$ be a bijection, and let $B = f(A)$. We can write

$$L_s\left((f^{-1}(B)\right) = f^{-1}(B). \tag{A.27}$$

We obtain

$$f\left(L_s\left((f^{-1}(B)\right)\right) = B, \tag{A.28}$$

which can be written as

$$f \circ L_s \circ f^{-1}(B) = B. \tag{A.29}$$

When the bijection f happens to be the action via some element $p \in \mathcal{G}$, namely when $f = L_p$, we conclude that the conjugate psp^{-1} is a symmetry of B. In the case of our ellipsoid, any transformation of the form

$$\{v\}' = [F][R][F]^{-1}\{v\}'', \tag{A.30}$$

where $[R]$ is an arbitrary rotation, leaves the ellipsoid unchanged. The group of symmetries of the ellipsoid is a conjugate of the proper orthogonal group via the transformation of the unit sphere.

A.3 Chapter 3

Exercise 3.1 To prove that a set $\mathcal{G}$ is a group we need first to define a binary operation and then to show (i) associativity, (ii) existence of a unit element, and (iii) existence of inverse elements. The operation in this case is (3×3)-matrix multiplication.

(i) Associativity: Let G_1 and G_2 be material symmetries at X of the function $W(\mathbf{F}, X)$. Then,

$$W(\mathbf{F}(\mathbf{G}_1\mathbf{G}_2), X) = W((\mathbf{F}\mathbf{G}_1)\mathbf{G}_2, X) = W(\mathbf{F}(\mathbf{G}_1), X) = W(\mathbf{F}, X). \tag{A.31}$$

(ii) Unit element:

$$W(\mathbf{F}\mathbf{I}, X) = W(\mathbf{F}, X). \tag{A.32}$$

(iii) Inverses: Let $\mathbf{G}$ be a symmetry at X. Then, $\forall \mathbf{F} \in GL^{+}(3, \mathbb{R})$, we obtain

$$W(\mathbf{F}\mathbf{G}^{-1}, X) = W((\mathbf{F}\mathbf{G}^{-1})\mathbf{G}, X) = W((\mathbf{F}\mathbf{G}^{-1}\mathbf{G}), X) = W(\mathbf{F}, X). \tag{A.33}$$

Exercise 3.2 Recalling that for the verification of the principle of frame indifference the rotation $\mathbf{Q}$ is applied to the left of $\mathbf{F}$, we obtain

$$W((\mathbf{Q}\mathbf{F})^{T}(\mathbf{Q}\mathbf{F}), X) = W(\mathbf{F}^{T}\mathbf{Q}^{T}\mathbf{Q}\mathbf{F}, X) = W(\mathbf{F}^{T}\mathbf{F}, X), \tag{A.34}$$

where the property of orthogonal matrices (for which the transpose equals the inverse) has been exploited. The proof of necessity makes use of the *polar decomposition theorem* of linear algebra, which states that every non-singular square matrix can be uniquely decomposed into the product of an orthogonal matrix and a positive-definite symmetric matrix. The answer to the question about a constitutive dependence on the product $\mathbf{F}\mathbf{F}^{T}$ is negative as far as the principle of frame indifference is concerned. Such dependence, however, when interpreted in terms of material (rather than spatial) deformations $\mathbf{Q}$, correspond to the concept of material isotropy.

Exercise 3.3 If a deformation $\chi : \mathcal{B} \to \mathbb{R}^3$ were not injective, two distinct body points would come to occupy simultaneously the same position in space. That would be a case of *interpenetration of matter*, an interesting situation that we are explicitly excluding from this more standard formulation.

Exercise 3.5 Let $L_g(x) = L_g(y)$ for some $x, y \in A$. We may apply $L_{g^{-1}}$ to this equality and invoke the properties of composition and identity to obtain $x = y$, which proves that the action is injective. To prove that the action is surjective (onto) take an arbitrary element $z \in A$. For each $g \in \mathcal{G}$ we can write

$$z = L_e(z) = L_{gg^{-1}}(z) = L_g\left(L_{g^{-1}}(z)\right), \tag{A.35}$$

which means that each z in A is the image of the action L_g on some element of A. Finally, having proved bijectivity, the inverse of L_g is well-defined. If $y = L_g(x)$ we have $x = \left(L_g\right)^{-1}(y)$. But we also have that $L_{g^{-1}}(y) = x$, from where we obtain $\left(L_g\right)^{-1} = L_{g^{-1}}$.

Exercise 3.11 The reflections $[S]_x$ and $[S]_y$, respectively about the planes yz and xz, are represented by the matrices

$$[S]_x = \begin{bmatrix} -1 & 0 & 0 \\ 0 & 1 & 0 \\ 0 & 0 & 1 \end{bmatrix}, \qquad [S]_y = \begin{bmatrix} 1 & 0 & 0 \\ 0 & -1 & 0 \\ 0 & 0 & 1 \end{bmatrix}, \qquad [S]_z = \begin{bmatrix} 1 & 0 & 0 \\ 0 & 1 & 0 \\ 0 & 0 & -1 \end{bmatrix}. \tag{A.36}$$

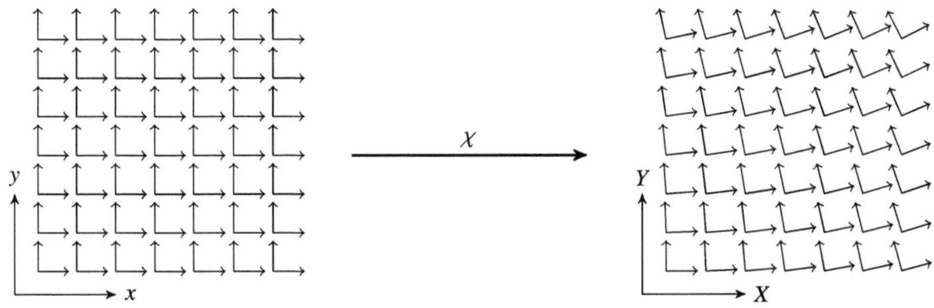

Fig. A.4 Mapping a homogeneous configuration

The remaining five elements of the group can be obtained as

(a)

$$= [-I][-I], \quad [S]_z = [-I][S]_x[S]_y, \quad [\hat{R}]_z = [S]_x[S]_y,$$
$$[\hat{R}]_y = [-I][S]_y, \quad [\hat{R}]_x = [-I][S]_x. \tag{A.37}$$

(b)

$$= [-I][-I], \quad [\hat{R}]_z = [\hat{R}]_x[\hat{R}]_y, \quad [S]_z = [-I][\hat{R}]_x[\hat{R}]_y,$$
$$[S]_y = [-I][\hat{R}]_y, \quad [S]_x = [-I][\hat{R}]_x. \tag{A.38}$$

Exercise 3.15 The body is homogeneous if, and only if, it can be brought by means of a smooth deformation to a homogeneous configuration, whereby the field of implants is constant. Equivalently, the body is homogeneous if, and only if, the given reference configuration is obtainable as a smooth deformation of a homogeneous configuration, as schematically shown in Fig. A.4.

Let the supposed deformation be given by

$$\chi = \begin{cases} X = X(x, y) \\ Y = Y(x, y) \end{cases}. \tag{A.39}$$

The gradient of this hypothetical deformation should be equal, at each point, to the given implant field $[P]$. Accordingly, we must have

$$[P] = \begin{bmatrix} \cos\theta(X, Y) & -\sin\theta(X, Y) \\ \sin\theta(X, Y) & \cos\theta(X, Y) \end{bmatrix} = \begin{bmatrix} \frac{\partial X}{\partial x} & \frac{\partial X}{\partial y} \\ \frac{\partial Y}{\partial x} & \frac{\partial Y}{\partial y} \end{bmatrix}. \tag{A.40}$$

Taking mixed derivatives, we obtain the compatibility conditions

$$\begin{aligned}\frac{\partial^2 X}{\partial x \partial y} &= -(\sin\theta)\frac{\partial\theta}{\partial y} = -(\cos\theta)\frac{\partial\theta}{\partial x},\\ \frac{\partial^2 Y}{\partial x \partial y} &= (\cos\theta)\frac{\partial\theta}{\partial y} = -(\sin\theta)\frac{\partial\theta}{\partial x}.\end{aligned} \tag{A.41}$$

These two conditions are mutually contradictory unless θ is constant.

More generally, let the matrix $[P]$ be given as a function of X and Y. The entries P_1^1, P_1^2 are the Cartesian components of the first base-vector, while the entries P_2^1, P_2^2 are the components of the second base-vector. For this basis to have been produced via a smooth deformation of a Cartesian basis, we must have

$$\begin{aligned}\frac{\partial X}{\partial x} &= P_1^1, & \frac{\partial Y}{\partial x} &= P_1^2,\\ \frac{\partial X}{\partial y} &= P_2^1, & \frac{\partial Y}{\partial y} &= P_2^2.\end{aligned} \tag{A.42}$$

It follows from these equations that a necessary condition for the basis field to be the image of a Cartesian vector field via a smooth deformation is given by the equality of the mixed second derivatives, namely,

$$\frac{\partial P_1^1}{\partial y} = \frac{\partial P_2^1}{\partial x}, \qquad \frac{\partial P_1^2}{\partial y} = \frac{\partial P_2^2}{\partial x}. \tag{A.43}$$

Since the functions in $[P]$ are given in terms of X and Y, we can used the chain rule and rewrite these conditions as

$$\begin{aligned}\frac{\partial P_1^1}{\partial X}P_2^1 + \frac{\partial P_1^1}{\partial Y}P_2^2 &= \frac{\partial P_2^1}{\partial X}P_1^1 + \frac{\partial P_2^1}{\partial Y}P_1^2,\\ \frac{\partial P_1^2}{\partial X}P_2^1 + \frac{\partial P_1^2}{\partial Y}P_2^2 &= \frac{\partial P_2^2}{\partial X}P_1^1 + \frac{\partial P_2^2}{\partial Y}P_1^2.\end{aligned} \tag{A.44}$$

These conditions are also sufficient, at least in an open neighbourhood of each point, for homogeneity. It is possible, since we have come this far in this digression, to generalize this result more elegantly for any number n of dimensions. Let us denote the entries in the matrix $[P]$ by P_α^I $(I, \alpha = 1, \ldots, n)$. The Cartesian coordinates in the reference configuration are X^I $(I = 1, \ldots, n)$ and the Cartesian coordinates in the putative homogeneous configuration by x^α $(\alpha = 1, \ldots, n)$. Adhering to Einstein's summation convention, whereby diagonally repeated indices in a monomial indicate summation over their range, and notating partial derivatives by means of a comma preceding an index, Eq. (A.40) can be compactly rewritten as

$$X^I_{,\alpha} = P^I_\alpha, \qquad I, \alpha = 1, \ldots, n. \tag{A.45}$$

By equality of mixed partials and application of the chain rule we obtain

$$P^K_{\alpha,J} P^J_\beta = P^K_{\beta,J} P^J_\alpha. \tag{A.46}$$

These are the required compatibility conditions for the body to be homogeneous. In terms of differential geometric concepts, the generally *non-holonomic frame* $[P](X^I)$ defines a *distant parallelism*. The associated *linear connection* is characterized by the *Christoffel symbols*

$$\Gamma^K_{IJ} = -P^K_{\alpha,J}\left(P^{-1}\right)^\alpha_I. \tag{A.47}$$

The *torsion tensor* of the connection is defined as the skew-symmetric tensor τ with components

$$\tau^K_{IJ} = \frac{1}{2}(\Gamma^K_{IJ} - \Gamma^K_{JI}). \tag{A.48}$$

In geometric terms, therefore, the homogeneity condition is expressed as the vanishing of the torsion tensor. Equivalently, we may say that the torsion tensor is a measure of the density of dislocations in the body. Finally, it can be said that the vanishing of the torsion guarantees (at least locally) the existence of a coordinate patch such that its natural basis coincides at each point of the domain with the given frame field.

A.4 Chapter 4

Exercise 4.2 We have four different groupoid structures in a double groupoid, namely, the two side groupoids $\mathcal{H} \rightrightarrows \mathcal{B}$ and $\mathcal{V} \rightrightarrows \mathcal{B}$, and the two groupoid structures $Q \rightrightarrows \mathcal{H}$ and $Q \rightrightarrows \mathcal{V}$. The set of objects of the horizontal structure are arrows in $\mathcal{V}$, and the 'arrows' are pairs of arrows in $\mathcal{H}$. The tail of one of these hyper-arrows is obtained via the projection $\tilde{\alpha}_H$. This tail is a vertical arrow. The tip is obtained by means of the projection $\tilde{\beta}_H$. To multiply horizontally two squares, such as $\mathbb{A} \odot_H \mathbb{A}'$ we need the tip of $\mathbb{A}'$ to coincide with the tail of $\mathbb{A}$, that is,

$$\tilde{\beta}_H(\mathbb{A}') = \tilde{\alpha}_H(\mathbb{A}). \tag{A.49}$$

Exercise 4.5 Recalling the definition of the (simple) pair groupoid in Example 1.3, a typical square of the pair double groupoid spanning four points of $\mathcal{B}$ can be

represented as

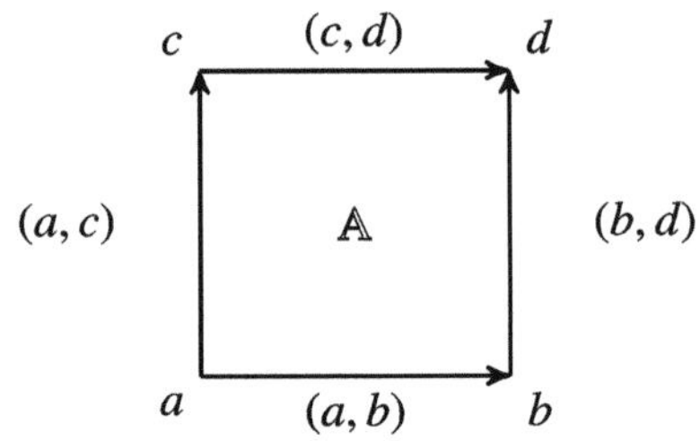

(A.50)

The projections deliver

$$\tilde{\alpha}_H(\mathbb{A}) = (a, c), \qquad \tilde{\beta}_H(\mathbb{A}) = (b, d), \tag{A.51}$$

and

$$\tilde{\alpha}_V(\mathbb{A}) = (a, b), \qquad \tilde{\beta}_V(\mathbb{A}) = (c, d). \tag{A.52}$$

Exercise 4.6 Use the commutativity of a pair of products at a time, that is,

$$\begin{aligned} &((\mathbb{A} \odot_I \mathbb{B}) \odot_J (\mathbb{C} \odot_I \mathbb{D})) \odot_K ((\mathbb{M} \odot_I \mathbb{N}) \odot_J (\mathbb{P} \odot_I \mathbb{Q})) \\ &= ((\mathbb{A} \odot_J \mathbb{C}) \odot_I (\mathbb{B} \odot_J \mathbb{D})) \odot_K (\mathbb{M} \odot_J \mathbb{P})) \odot_I ((\mathbb{N} \odot_J \mathbb{Q})) \\ &= ((\mathbb{A} \odot_J \mathbb{C}) \odot_K (\mathbb{M} \odot_J \mathbb{P})) \odot_I (\mathbb{B} \odot_J \mathbb{D})) \odot_K ((\mathbb{N} \odot_J \mathbb{Q})) \\ &= ((\mathbb{A} \odot_K \mathbb{M}) \odot_J (\mathbb{C} \odot_K \mathbb{P})) \odot_I (\mathbb{B} \odot_K \mathbb{N})) \odot_J ((\mathbb{D} \odot_K \mathbb{Q})) . \end{aligned} \tag{A.53}$$

A.5 Chapter 5

Exercise 5.1 Figure A.5 is a reproduction of Fig. 5.2 with the addition of the proper orthogonal matrices $\mathbf{G}$ and $\hat{\mathbf{G}}$ representing the degrees of freedom in the choice of implants from the respective stress-free archetypes.

If the composite is uniform we can find (for each pair of points X_1 and X_2) two proper orthogonal matrices $\mathbf{G}$ and $\hat{\mathbf{G}}$ such that

$$\mathbf{P}(X_2)\,\mathbf{G}\,(\mathbf{P}(X_1))^{-1} = \hat{\mathbf{P}}(X_2)\,\hat{\mathbf{G}}\left(\hat{\mathbf{P}}(X_1)\right)^{-1}. \tag{A.54}$$

Recalling the definition $\mathbf{A}(X) = \left(\hat{\mathbf{P}}(X)\right)^{-1}\mathbf{P}(X)$, we can rewrite this condition as

$$\mathbf{A}(X_2)\,\mathbf{G} = \hat{\mathbf{G}}\hat{\mathbf{A}}(X_1). \tag{A.55}$$

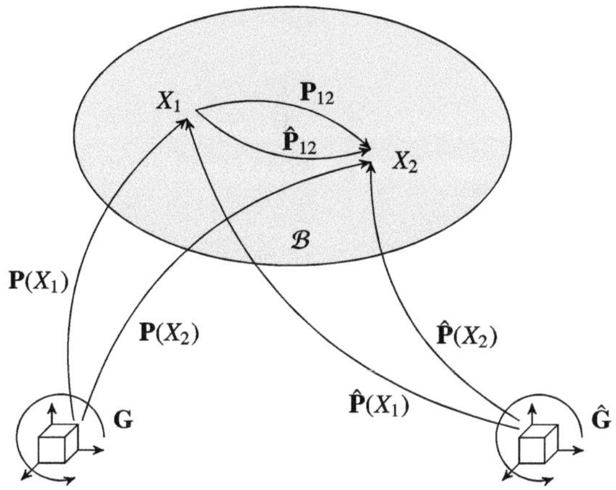

Fig. A.5 A binary composite made of two uniform isotropic constituents

Multiplying each side to the left by its transpose, we obtain

$$\mathbf{G}^T\mathbf{B}(X_2)\,\mathbf{G} = \mathbf{B}(X_1), \tag{A.56}$$

where $\mathbf{B}(X) = \mathbf{A}^T(X)\,\mathbf{A}(X)$ is a positive-definite symmetric field, already introduced in Eq. (5.8). In other words, the matrices $\mathbf{B}(X_1)$ and $\mathbf{B}(X_2)$ are *orthogonally similar*. We conclude that a necessary condition for uniformity of the composite is that the field $\mathbf{B}(X)$ must have constant (and, incidentally, positive) eigenvalues. Conversely, let the symmetric field $\mathbf{B}(X)$ have constant eigenvalues. This would imply that, for any pair X_1, X_2, there exists an orthogonal matrix $\mathbf{G}$ such that

$$\mathbf{G}^T\mathbf{A}^T(X_2)\mathbf{A}(X_2)\,\mathbf{G} = \mathbf{A}^T(X_1)\mathbf{A}(X_1), \tag{A.57}$$

which implies that

$$((\mathbf{A}(X_2))\mathbf{G}\mathbf{A}(X_1))^T = ((\mathbf{A}(X_2))\mathbf{G}\mathbf{A}(X_1))^{-1}. \tag{A.58}$$

Hence, the matrix $((\mathbf{A}(X_2))\mathbf{G}\mathbf{A}(X_1))$ is orthogonal, and we recover Eq. (A.55).

Exercise 5.3 Since the symmetry groups of both constituents are discrete, we use the tensor field $\mathbf{A}$ as the measure of misalignment. A fully aligned (i.e., uniform) composite would correspond to a constant field $\mathbf{A}$. In our case we obtain

$$[A] = [\hat{P}]^{-1}[P] = \begin{bmatrix} \cos\theta(x,y) & \sin\theta(x,y) \\ -\sin\theta(x,y) & \cos\theta(x,y) \end{bmatrix}, \tag{A.59}$$

where we have exploited the orthogonality of the matrix $[\hat{P}]$ and the fact that $[P] = [I]$. Since $\theta(x,y)$ is not constant, neither is $[A]$, which implies that the composite is (intentionally) non-uniform. The local measure of non-uniformity can be identified with the gradient of $[A]$, which would be a third-order Cartesian tensor.

Its components can be represented as the derivative matrices $[A]_{,x}$ and $[A]_{,y}$ given by

$$[A]_{,x} = \begin{bmatrix} -\sin\theta(x, y) & \cos\theta(x, y) \\ -\cos\theta(x, y) & -\sin\theta(x, y) \end{bmatrix} \frac{\pi}{4} \sin(\pi x) \tag{A.60}$$

and

$$[A]_{,y} = \begin{bmatrix} -\sin\theta(x, y) & \cos\theta(x, y) \\ -\cos\theta(x, y) & -\sin\theta(x, y) \end{bmatrix} \frac{3\pi}{10} \sin(2\pi y). \tag{A.61}$$

A.6 Chapter 6

Exercise 6.3

(a) The weight of the walk is

$$[W] = [D]^{-1}[C]^{-1}[B][A] = \begin{bmatrix} 0 & 0 & 1 \\ 0 & -1 & 0 \\ 2 & 0 & 0 \end{bmatrix} \tag{A.62}$$

(b) For the second trajectory, the weight is given by

$$[W]' = [A]^{-1}[B]^{-1}[A]^{-1}[B][B][D]^{-1}[C][D] = \begin{bmatrix} 1/2 & 0 & 0 \\ 0 & -1 & 0 \\ 0 & 0 & -1/2 \end{bmatrix}. \tag{A.63}$$

Exercise 6.4

(a) The weight of the walk is

$$[W] = -[D] - [C] + [B] + [A] = \begin{bmatrix} 2 & 0 & -3 \\ 0 & 2 & 0 \\ 1 & 0 & 1 \end{bmatrix} \tag{A.64}$$

(b) For the second trajectory, the weight is given by

$$[W]' = -[A] - [B] - [A] + [B] + [B] - [D] + [C] + [D] = \begin{bmatrix} -4 & 0 & 0 \\ 0 & -2 & 0 \\ 2 & 0 & -2 \end{bmatrix}. \tag{A.65}$$

Exercise 6.7 Let us introduce the following notation to designate the various sub-mixtures: for a mixture with n basic simple constituents, the label of each sub-mixture consists of a sequence of n spaces, one for each basic constituent. Each space is filled with either a 0 or a 1, according to whether the corresponding basic constituent is absent or present in the sub-mixture. The total mixture, therefore, consists of a sequence of n 1's, whereas the empty sub-mixture is a sequence of n 0's. With the arrows defined as stipulated in the statement of the exercise, we obtain a one-to-one correspondence between this graph and the graph of an n-skeleton, such as the one represented in Fig. 6.8.

A.7 Chapter 7

Exercise 7.1

The pattern can be described as follows: the edges of class X_I run between nodes whose numbers differ by 2^{n-I}, where n is the dimension. In the case of the tesseract ($n = 4$) those differences are respectively $8, 4, 2, 1$ for the edges parallel to the coordinates X_1, X_2, X_3, X_4 (Fig. A.6).

Exercise 7.5 For the case $n = 4$, Eq. (7.3) delivers

$$P_4 = (4)\, 2^3 = 32. \tag{A.66}$$

The corresponding hierarchical structure, plotted in Fig. 7.3, shows 32 double links. Similarly, for the case $n = 3$, we obtain

$$P_3 = (3)\, 2^2 = 12, \tag{A.67}$$

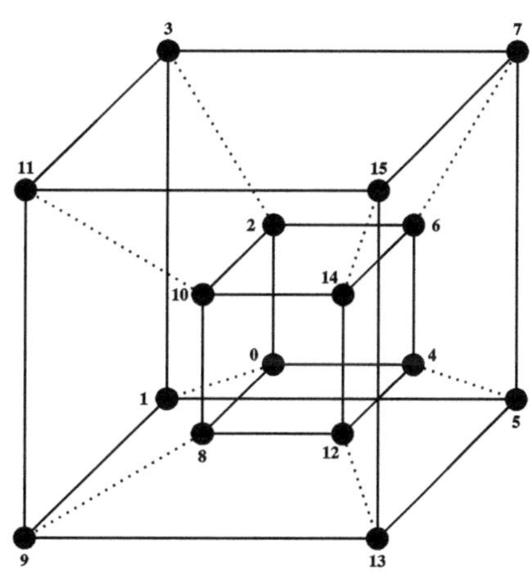

Fig. A.6 Decimal numbering of the vertices of a tesseract

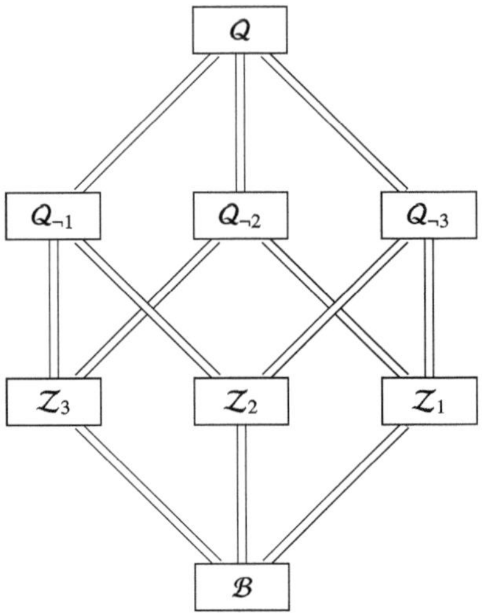

Fig. A.7 Hierarchical structure of the projection maps for $n = 3$

which is the number of double links in the structure depicted in Fig. A.7. Notice how this diagram can be also regarded as the projection on the plane of a cube, whose edges happen to represent the various groupoid projection maps.

Exercise 7.9 In the case of a trivial symmetry group (triclinic material), alignment between constituents is an equivalence relation. Reflexivity and symmetry are obvious. To prove transitivity, consider the frame fields $\mathbf{P}_1(X)$, $\mathbf{P}_2(X)$, and $\mathbf{P}_3(X)$, associated with three constituents. If constituent 2 is aligned with constituent 1, Eq. (5.3) implies that

$$\mathbf{P}_1(X) = \mathbf{P}_2(X)\,\mathbf{K}, \tag{A.68}$$

where $\mathbf{K}$ is a constant matrix. Similarly, if constituent 3 is aligned with constituent 2 it follows that

$$\mathbf{P}_2(X) = \mathbf{P}_3(X)\,\mathbf{L}, \tag{A.69}$$

where $\mathbf{L}$ is a constant matrix. Combining these two equations we obtain

$$\mathbf{P}_1(X) = \mathbf{P}_3(X)\,\mathbf{L}\,\mathbf{K}, \tag{A.70}$$

which proves the assertion. Intuitively, for the case of trivial symmetry, the alignment between two components implies that they share the only arrow at their disposal between any two given points. Transitivity follows from the fact that no other arrow is available for alignment with other components.

References

1. Akhmetshin, L., Iokhim, K., Kazantseva, E., Smolin, I.: Influence of topological defects on the mechanical response of unit cells of the tetrachiral mechanical metamaterial. Designs **7**/129 (2023). https://doi.org/10.3390/designs7060129
2. Andruskiewitsch, N., Natale, S.: The structure of double groupoids. J. Pure Appl. Algebra **213**, 1031–1045 (2009)
3. Askari, M.A., Nazari, M.A., Perrier, P., Payan, Y.: Evaluation of residual stresses in human face as a function of growth. Int. J. Comput. Electr. Automation Control Inf. Eng. **10**(1), 142–148 (2016)
4. Brandt, H.: Über eine Verallgemeinerung des Gruppenbegriffes. Math. Ann. **96**, 360–366 (1927)
5. Brown, R.: Topology and Groupoids (2006). Printed and distributed by Createspace, Book-Surge Publishing (retitled, revised, updated and extended edition of a classic text, originally published by McGraw-Hill in 1968)
6. Brown, R., Mackenzie, K.C.H.: Determination of a double Lie groupoid by its core diagram. J. Pure Appl. Algebra **80**, 237–272 (1992)
7. Coleman, B.D.: Simple liquid crystals. Arch. Ration. Mech. Anal. **20**, 41–58 (1965)
8. de León, M., Epstein, M., Jiménez, V.: Material Geometry. World Scientific, Singapore (2021)
9. Ehresmann, C.: Catégories structurées. Ann. Sci. l'École Normale Supérieure **80**(4), 349–426 (1963)
10. Epstein, M.: Material geometry of binary composites. Symmetry **13**, 892 (2021). https://doi.org/10.3390/sym13050892
11. Epstein, M.: The theory of continuous distributions of composite defects. J. Mech. Phys. Solids **167**, 104983 (2022)
12. Epstein, M., de León, M.: Homogeneity without uniformity: toward a mathematical theory of functionally graded materials. Int. J. Solids Struct. **37**, 7577–7591 (2000)
13. Epstein, M., Elżanowski, M.: Material Inhomogeneities and their Evolution: a Geometric Approach. Springer, Berlin (2007)
14. Feng, P., Wu, P., Gao, C., Yang, Y., Guo, W., Yang, W., Shuai, C.: A multimaterial scaffold with tunable properties: toward bone tissue repair. Adv. Sci. **5**, 1799817 (2018)
15. Fung, Y.C.: What are residual stresses doing in our blood vessels? Ann. Biomed. Eng. **19**, 237–249 (1991)
16. Grandis, M.: Higher Dimensional Categories: From Double to Multiple Categories. World Scientific, Singapore (2019)
17. Harary, F.: Graph Theory. Addison-Wesley, Reading (1969)

© The Author(s), under exclusive license to Springer Nature Singapore Pte Ltd. 2026
M. Epstein, *An Elementary Introduction to* n*-fold Groupoids*,
Surveys and Tutorials in the Applied Mathematical Sciences 19,
https://doi.org/10.1007/978-981-95-1869-2

18. Harary, F., Hayes, J.P., Wu, H.J.: A survey of the theory of hypercube graphs. Comput. Math. Appl. **15**(4), 277–289 (1988)
19. Henderson, J., Arya, R., Gillespie, P.: Skin graft meshing, over-meshing and cross-meshing. Int. J. Surgery **10**(9), 547–550 (2012)
20. Ibort, A., Rodríguez, M.A.: An Introduction to Groups, Groupoids and Their Representations. CRC Press/Taylor & Francis Group, Boca Raton/Abingdon (2021)
21. Isaacson, W.: Leonardo da Vinci. Simon & Schuster, New York (2017)
22. Mackenzie, K.: Lie Groupoids and Lie Algebroids in Differential Geometry. London Mathematical Society Lecture Note Series, vol. 124. Cambridge University Press, Cambridge (1987)
23. Mackenzie, K.C.H.: Double Lie algebroids and second-order geometry, I. Adv. Math. **94**(2), 180–239 (1992)
24. Mackenzie, K.C.H.: General Theory of Lie Groupoids and Lie Algebroids. London Mathematical Society Lecture Note Series, vol. 213. Cambridge University Press, Cambridge (2005)
25. Meeussen, A., Oğuz, E.C., Shokef, Y., van Hecke, M.: Topological defects produce exotic mechanics in complex metamaterials. Nature Phys. Lett. (2024). https://doi.org/10.1038/s41567-019-0763-6
26. Misner, C.W., Thorne, K.S., Wheeler, J.A.: Gravitation. W.H. Freeman and Company, San Francisco (1973)
27. Nazir, A., Gokcekaya, O., Masum Billah, K.M., Ertugul, O., Jiang, J., Sun, J., Hussain, S.: Multi-material additive manufacturing: a systematic review of design, properties, applications, challenges, and 3D printing of materials and cellular metamaterials. Mater. Design **226**, 111661 (2023)
28. Noll, W.: Proof of the maximality of the orthogonal group in the unimodular group. Arch. Ration. Mech. Anal. **18**, 100–102 (1965)
29. Noll, W.: Materially uniform bodies with inhomogeneities. Arch. Ration. Mech. Anal. **27**, 1–32 (1967)
30. Truesdell, C., Noll, W.: The Non-linear Field Theories of Mechanics. In: Flügge, S. (ed.) Encyclopedia of Physics, vol. III/3. Springer-Verlag, Berlin (1965)
31. Wang, C.-C.: A general theory of subfluids. Arch. Ration. Mech. Anal. **20**, 1–40 (1965)
32. Wang, C.-C.: On the geometric structure of simple bodies, a mathematical foundation for the theory of continuous distributions of dislocations. Arch. Ration. Mech. Anal. **27**, 33–94 (1967)
33. Wang, C., Vangelatos, Z., Grigoropoulos, C.P., Ma, Z.: Micro-engineered architected metamaterials for cell and tissue engineering. Mater. Today Adv. **13**, 100226 (2022)
34. Weinstein, A.: Groupoids: unifying internal and external symmetry. A tour through examples. Not. Am. Math. Soc. **43**(7), 744–752 (2000)
35. Yavas, D., Liu, Q., Zhang, Z., Wu, D.: Design and fabrication of architected multi-material lattices with tunable stiffness, strength, and energy absorption. Mater. Design **217**, 110613 (2022)
36. Yuan, C., Wang, F., Qi, B., Ding, Z., Rosen, D.W., Ge, Q.: 3D printing of multi-material composites with tunable shape memory behavior. Mater. Design **193**, 108785 (2020)
37. Zahouani, H., Ayadh, M., Abellan, M.A., Bigouret, A.: Considering the effect of residual tension forces on the wavelength anisometry of skin imaging by 2D skin tension integrity model. Sci. Rep. **14**, 31963 (2024)

Index

© The Author(s), under exclusive license to Springer Nature Singapore Pte Ltd. 2026
M. Epstein, *An Elementary Introduction to* n*-fold Groupoids*,
Surveys and Tutorials in the Applied Mathematical Sciences 19,
https://doi.org/10.1007/978-981-95-1869-2

MIX
Papier aus verantwortungsvollen Quellen
Paper from responsible sources
FSC® C105338

If you have any concerns about our products,
you can contact us on
ProductSafety@springernature.com

In case Publisher is established outside the EU,
the EU authorized representative is:
Springer Nature Customer Service Center GmbH
Europaplatz 3, 69115 Heidelberg, Germany

Printed by Libri Plureos GmbH
in Hamburg, Germany